D0207679

ANTITRUST AND THE OIL MONOPOLY

ANTITRUST AND THE OIL MONOPOLY

THE STANDARD OIL CASES, 1890-1911

Bruce Bringhurst

CONTRIBUTIONS IN LEGAL STUDIES, NUMBER 8

GREENWOOD PRESS
WESTPORT, CONNECTICUT • LONDON, ENGLAND

Library of Congress Cataloging in Publication Data

Bringhurst, Bruce.
　Antitrust and the oil monopoly.

　(Contributions in legal studies; no. 8 ISSN 0147-1074)
　Bibliography: p.
　Includes index.
　1. Standard Oil Company.　2. Antitrust law—
United States.　3. Petroleum law and legislation—
United States.　I. Title.　II. Series.
KF1866.S73B74　　　　343′.73′072　　　　78-67908
ISBN　0-313-20642-2

Library of Congress Catalog Card Number: 78-67908
ISBN: 0-313-20642-2
ISSN: 0147-1074

First published in 1979

Greenwood Press, Inc.
51 Riverside Avenue, Westport, Connecticut 06880

Printed in the United States of America

10 9 8 7 6 5 4 3 2 1

To Cheryl

Contents

Preface

Students of American politics are accustomed to a high level of governmental contradiction and deception; still there are some aspects of the political system that retain their ability to surprise the inadequately cynical observer. The role of the antitrust laws of the United States is a case in point. The American economy is dominated by large corporations that have organized themselves into cooperative oligopolies and have extended their reach far beyond the borders of the United States. The spectacular rise of the American-based multinationals, so often enemies of meaningful competition, has occurred since the passage of state and federal antitrust legislation in the late nineteenth century. Hence it is reasonable to conclude that these laws have been singularly ineffective in halting the ongoing concentration of American business. Nevertheless politicians, public interest advocates, and leading scholars repeatedly turn to the antitrust laws as a means of coping with problems created by corporate excess, and the federal government routinely pours resources into the antitrust division of the Justice Department.

This book is an attempt to uncover the reasons behind the failure of antitrust legislation to preserve competition and limit corporate power. It also explores the basis of the politicians' enduring fascination with these laws. The method employed is a detailed investigation of one of the key antitrust episodes in American history: the state and federal effort to control the Standard Oil combination between 1890 and 1911. This campaign is particularly significant because the petroleum industry has been one of the most critical and least controlled sectors of the American economy throughout the twentieth century. These suits are also important because they occurred during the first generation of antitrust litigation, when the laws were untested and American industry was predominantly national and therefore still within the jurisdiction of the federal government. Focusing on a single series of cases permits an examination of the motives behind a particular suit as well as its impact on corporate operations, and in antitrust litigation the political and economic context is frequently far more important than the legal arguments or the ruling of the court.

I wish to thank Professor Leonard W. Levy of the Claremont Graduate School, who suggested this topic, read the work in its initial stages, and provided a model of scholarly commitment. A grant from the John Randolph Haynes and Dora Haynes Foundation of Los Angeles financed a substantial amount of the research. Finally, my wife, Cheryl, participated in this effort in a wide variety of ways and made the completion of this frequently discouraging project possible.

Bruce Bringhurst

ANTITRUST AND THE OIL MONOPOLY

Introduction:
The Antitrust Dilemma

Americans in the twentieth century have grown accustomed to an economic, social, and political system dominated by a relative handful of giant corporations. But these organizations did not attain their present position of power and privilege without experiencing periodic outbursts of public hostility. In fact, concern over the concentration of economic power is an old American tradition. In the late 1880s, this concern emerged as a major political issue. The press lavished attention on the problem of monopoly in those years. Economists and political philosophers pondered the implications of the demise of unrestricted competition. And the issue became intimately connected with other leading political concerns of the day, such as tariff reform, the chronic farm problem, and urban poverty. In the presidential campaign of 1888, both Republicans and Democrats included passionate antimonopoly declarations in their party platforms.[1]

This public controversy resulted from the wave of combination that swept American industry during the 1880s. The most notable consolidation of the decade occurred in 1882 when John D. Rockefeller and his associates created the Standard Oil Trust. Capitalized at $70 million and controlling 90 percent of the rapidly growing petroleum industry, Standard Oil was an awesome economic power from the moment of its inception.[2] Fueled by enormous profits, the Standard organization expanded rapidly. Today the oil trust's major descendant, Exxon, is the largest industrial corporation in the world, with assets over $36 billion and 1976 sales of over $48 billion. Three other Standard Oil off-spring—Mobil Oil, Standard Oil of California (Chevron), and Standard Oil of Indiana (Amoco)—are currently included in the top ten corporations in the United States. Several other companies with origins in the Rockefeller trust are also among America's leading corporate powers, including Atlantic Richfield (Arco), Continental Oil (Conoco), and Standard Oil of Ohio (Sohio).[3]

The startling success of the Standard Oil trust agreement encouraged others to seek monopoly. In 1887, fourteen corporations merged to form the sugar trust, which was capitalized at $50 million and controlled 70 percent of American sugar-refining capacity. That same year the owners of eighty-one companies created a whiskey trust that manufactured at least 85 percent of the nation's alcohol and spirits. Despite these formidable consolidations, monopoly remained the exception. The American economy was expanding so rapidly and industrial technology was changing so quickly that very few of the trusts were actually capable of duplicating Standard Oil's control of an entire industry. Yet there could be no question about the size and power of the new combinations and the resulting public concern. In a fundamental sense, it was the vast economic power of the men who ran these new giants that posed the greatest threat to American institutions. Wealth is easily translated into political power, and the effective exercise of democratic government becomes increasingly difficult as control over society's economic resources concentrates in ever fewer hands.[4]

Politicians turned to the law in their attempt to deal with the trust controversy. The common law expressed the traditional English and American antipathy toward monopoly, and most lawyers believed that it was a potentially useful tool. The common law also condemned government-chartered monopolies, as well as the antiquated practices of engrossing, regrating, and forestalling. Unfortunately, these branches of the law were of little use in dealing with modern industrial combinations. But the common law also prohibited contracts and combinations in restraint of trade, and these precedents had been occasionally employed to break up monopolies. More important than its modest actual attainments, the common law expressed a public policy that condemned monopoly.

As the Ohio Supreme Court declared in 1880, "Public policy, unquestionably, favors competition in trade, to the end that its commodities may be afforded to the customer as cheaply as possible, and is opposed to monopolies, which tend to advance market prices, to the injury of the general public."[5]

Most opponents of monopoly thought that statutory prohibitions were required to supplement the common law. Reformers filed at least three antimonopoly bills in every state legislature in the 1880s. Between March 1889 and July 1890, thirteen states passed antitrust laws, and fourteen others enacted similar measures by the turn of the century. Although they varied considerably in detail, these state antitrust laws were broadly similar in intent. They made contracts and combinations, which were merely unenforceable as common law, illegal. They also attempted to transform the general common-law declarations against monopoly and restraint of trade into specific prohibitions against certain business practices, such as pooling, price controls, regulating or limiting production, local price discrimination, restraint of trade, exclusive contracts, agreements to divide operating territories, and agreements not to deal with certain persons or firms.[6]

The state attorney general was customarily responsible for enforcement of the antitrust law, although in a number of states he shared that duty with county or district attorneys. The laws frequently provided for personal fines and imprisonments for the officers of trusts engaged in illegal activities, but the primary penalties were stiff corporate fines and the revocation of corporate charters and business permits. The laws of many states also allowed private parties to sue to recover damages inflicted by firms engaged in illegal business practices.[7]

Although state antitrust laws could conceivably protect local competition, the new industrial combinations operated nationally. Thus national legislation was the only realistic solution to the larger monopoly problem. Senator John Sherman of Ohio led the effort to secure such a law. In 1889 he submitted an antitrust bill to the Senate of the Fifty-first Congress.[8] Conservative Republicans dominated this "billion dollar" Congress. On the surface they had little reason to support an antitrust act, but the Republicans wanted to raise tariff duties, and the Democrats long had charged that high tariffs bred trusts. In order to place some distance between their party and the monopolies, the Republican leadership was willing to support an antitrust law. Moreover, some Republicans were deeply disturbed by the public uproar over industrial combination and the resulting economic inequality.[9]

The Senate Judiciary Committee, which drafted the final language of the antitrust bill, produced a measure notable primarily for its ambiguities. It had two major substantive provisions:

Section 1. Every contract, combination in the form of trust or otherwise, or conspiracy, in restraint of trade or commerce among the several states, or with foreign nations, is hereby declared to be illegal.
Section 2. Every person who shall monopolize, or attempt to monopolize, or combine or conspire with any other person or persons, to monopolize any part of the trade or commerce among the several States or with foreign nations, shall be guilty of a misdemeanor.[10]

Based on the congressional power to regulate interstate commerce, this language did not directly attack trusts. It merely outlawed combinations that intended to restrict interstate shipment, and intent was always difficult to prove. Moreover, the committee could not settle the question of whether a manufacturing combination that subsequently transported goods in interstate commerce was in violation of the law.

During the debate on the committee bill, several senators requested clarification of the measure's sweeping language. Did it embrace literally every contract or combination in restraint of trade? Such an interpretation might include all business contracts. A member of the Judiciary Committee attempted an explanation: "We have affirmed the old doctrine of the common law in regard to all interstate and international commercial transactions, and have clothed the United States courts with authority to enforce that doctrine by injunction."[11] This reference to the common law did little to clarify the scope of the act because the precedents on the matter were notoriously vague. But the lawmakers were apparently more interested in action than clarity. The Senate passed the bill by a vote of fifty-two to one, and the House enacted the measure unanimously. President Benjamin Harrison signed the Sherman Act into law on July 2, 1890.[12]

The enforcement provisions of the new act granted the United States circuit courts jurisdiction in antitrust cases and directed the attorney general to institute proceedings to "prevent and restrain" violations. Corporate officials convicted under the act were subject to fines as high as $5,000 and imprisonment for as long as one year for each infraction. The law provided for forfeiture of property shipped across state lines by an individual or corporation operating illegally. Finally, the act provided that a private citizen whose business or property was harmed by an illegal trust could file suit in a United States circuit court to recover three times the damages that he had sustained plus court costs.[13]

From the beginning, procedural problems plagued would-be trustbusters. A relative handful of prosecutors had to conduct reasonably thorough investigations of the business practices of large and complex organizations. They confronted various forms of corporate obstructionism while gathering evidence: corporate officials frequently refused to testify or provide material, delayed investigations by continuous consultations with their attorneys, or claimed that the required witness was out of town or that the relevant books had been lost or destroyed. In the end, the company under investigation might provide a mountain of material, much of it irrelevant, that required a great deal of effort to analyze. Whatever the form of resistance, the result was an incredibly time-consuming investigation.[14]

Once a suit came to trial, the prosecution confronted a host of new difficulties. A complex problem of industrial organization had to be molded into the form of a legal dispute. Legal procedure and the rules of evidence became paramount. Economic issues—the substance of the controversy—necessarily assumed a secondary role. Teams of corporation lawyers, always well paid and frequently brilliant, employed a bewildering variety of legal tactics to obscure and delay the entire process. In the end, judges with little or no expertise in economic matters made the final determination of guilt or innocence. If the company lost, it possessed ample resources to appeal to a higher court, frequently to several higher courts—a process that consumed years.

The prosecution's difficulties did not end with a conviction by the highest appellate court. A state court could impose only remedies narrowly limited by state borders. A federal decree in equity also had serious inherent defects. The decree did nothing to disturb the profits gained from past illegal activities. The defendant remained free to pursue the same general goals as before: the pursuit of profit and an enhanced share of the market. Moreover, the terms of the decree were static: prohibitions against specified arrangements or practices. But business conditions were fluid, and new opportunities to evade the basic purpose of the decree, without violating its provisions, rapidly emerged. The main problem with court-imposed remedies, however, was enforcement. Courts lacked the power to ensure compliance with their directives. If a law enforcement agency discovered a violation, it would have to initiate a new suit, a process that involved all the difficulties that plagued the initial effort.

Behind the procedural difficulties, there was a more fundamental problem with antitrust litigation. By the time that Congress passed the Sherman Act in 1890, large corporations had achieved a dominant position in the American economy. Economic hegemony allowed these giant enterprises to exert great influence on political, social, and ideological trends in the United States. Because antitrust litigation was so complex, customarily requiring years for the completion of a case, the defendant corporation had ample opportunity to exert its influence on the legal proceedings in a variety of ways.

In the early years of the antimonopoly movement, major corporations were not above a direct application of cash to derail an antitrust suit. The trusts hired leading regional and national politicians, ostensibly as legal advisers, to help cool the zeal of antimonopolists or mitigate the effects of antitrust litigation. The combination frequently employed lobbyists who attempted to discourage officials from resorting to lawsuits. Businessmen also employed public relations specialists to defuse public hostility and thus avoid prosecution for anticompetitive practices. Finally, massive campaign contributions by the corporations to political parties at both the state and federal levels gave those organizations great political leverage that proved very valuable in fending off antitrust suits.

The trusts used their economic position in another less direct fashion. The major combinations were important to the national economy, but they were often the very foundation of several local economies where they provided large numbers of jobs and paid substantial taxes. Hence a combination could apply economic pressure on a state government when faced with an antitrust challenge. Great enterprises could afford to offer incentives to states to drop lawsuits. Corporations lowered the prices of their products to dissipate antimonopoly sentiment. Corporate leaders also promised to make substantial investments in areas that demonstrated willingness to work with, rather than against, big business. Conversely, the combinations possessed the power to disrupt a local economy. The discontinuation of a major construction project or a substantial layoff at a factory were powerful arguments against pressing forward with an antitrust suit.

The attitude of political leaders, prosecutors, and judges concerning fundamental economic issues such as the sanctity of private property, the need for industrial development, and the desirability of big business was another critical factor in determining the outcome of antitrust cases. The vast majority of those associated with antitrust litigation had profound respect for property rights. This attitude frequently appeared in judicial decisions forcefully condemning certain corporate practices, and it tempered judgments against combinations on many occasions. At some point in this type of case, the prosecutor, who successfully had surmounted the various obstacles erected by the defendant corporation, and the judge who had to render final judgment, confronted a dilemma: should they insist on strict application of the law at the cost of disrupting the economy and limiting property rights? Although some stood firm, invariably a prosecutor or judge had second thoughts, and the defendant combination was alert to any sign of irresolution. In the end, corporate officials usually discovered a means of escape.

With all these problems, the antitrust laws proved totally incapable of halting the ongoing concentration of American industry. It is therefore not surprising that both contemporaries and later historians charged that Congress drafted an intentionally vague law that would satisfy the public without harming the trusts. Senator Orville Pratt of Connecticut, for example, bluntly declared during the debate on the Sherman Act, "The conduct of the

Senate . . . has not been in the line of the honest preparation of a bill to prohibit and punish trusts. It has been in the line of getting some bill with that title that we might go to the country with."[15] But the most thorough analyses of the actual framing of the Sherman Act have failed to uncover any such sinister plot to dupe the public. Moreover, any simple conspiracy theory failed to explain the ineffectuality of the numerous state antitrust laws. The weakness of all these laws apparently resulted from the inveterate conservatism of most legislators, a desire to write broad statutes of general applicability based on proven—if somewhat obscure —common-law precedents, and the newness and complexity of the trust problem.[16]

Although it failed to restructure the economy, the antitrust movement eventually established itself as a fixture in the American political system. After World War II it emerged as an important instrument of technical economic adjustment administered by specialists and benefiting the corporations at least as much as the public. But technical adjustment was not what the early trustbusters had in mind, and the Standard Oil cases—the main focus here—were not determined by court decisions based upon the pertinent economic facts at issue. Hence readers familiar with contemporary antitrust litigation probably will be surprised at the relatively minor emphasis placed on economic analysis. The evolution of the suits against Standard Oil between 1890 and 1911 was primarily dependent on legal technicality, political maneuver, and press manipulation. These cases frequently sprang from legal or political complications and rarely had a significant economic impact. The abuses of Standard Oil were very real, and the investigations that resulted from these suits produced a wealth of data that underscored that fact. But as it happened, the suits themselves were not intimately connected with that underlying economic reality.[17]

Over the years the antitrust laws assumed another, perhaps even more important, function. Since the 1880s many Americans have been deeply disturbed by the emergence of new combinations that have rapidly changed the economic foundation of their lives, and they have demanded political action to halt the spread of monopoly. Although these laws provided little real opportunity to change the shape of the economy, they did offer a mechanism through which the political system could act out its endlessly professed devotion to the classic capitalism of small-scale enterprise and unrestricted competition. This does not mean that antitrust prosecutors were routinely cynical individuals with few illusions and hidden motives. It is merely a recognition of the fact that the actual long-term impact of a particular legal or political process does not necessarily mirror the intentions of those who either initiate or perpetuate that process. Thus antitrust litigation served the highly important, although largely unintended, function of defusing the periodic outbursts of public hostility directed against the ongoing concentration of economic power in America. The Standard Oil cases mark the coming of age of antitrust as an important mechanism for coping with public outrage. No sector of the economy has benefited more from that service than the petroleum industry.

(1)
Ohio versus Standard Oil

The Standard Oil trust originated in Cleveland in 1863 when John D. Rockefeller, then twenty-six years old and grossing a comparatively pitiful $16,000 a year dealing in commodities, invested in Samuel Andrew's kerosene refinery. At that time Cleveland was the nation's leading oil-refining center. The city was near the source of crude oil in western Pennsylvania, as well as the rapidly growing midwestern market. Several competing railroad lines offered cheap transportation, and the Great Lakes and the Erie Canal provided an alternate means of shipping petroleum products to market. Blessed with advantageous location, clever management, and the rapid growth of the oil business, the enterprise prospered. William Rockefeller, John's brother, soon joined the company, serving as sales agent in New York. And Henry M. Flagler, who exhibited a remarkable talent for extracting bargain rates from railroads, became a managing partner. In 1870 these men created the Standard Oil Company of Ohio. Capitalized at $1 million, the new corporation already controlled 10 percent of the nation's oil-refining capacity.[1]

Rockefeller and his associates shrewdly avoided oil production because producers were constantly plagued by surplus crude and low prices. Instead they concentrated on refining and distributing petroleum products, and they employed a variety of tactics to gain control over those sectors of the industry. The Standard leaders actively participated in central agencies, through which all refiners could purchase crude, negotiate railroad rates, and market refined products. More important, they used their position as a leading shipper to extract special rates and rebates from the railroads. Ultimately they absorbed most of their competitors by offering them either an interest in Standard Oil or a cash settlement, while threatening to drive them out of business if they did not accept the offer. These methods proved spectacularly successful. By 1880 the Standard organization directly owned or effectively controlled 90 to 95 percent of the nation's petroleum refining and transport facilities. Overwhelming dominance in these key sectors of the industry gave Standard Oil great influence over both crude oil production and refined product marketing.[2]

This massive organization presented Rockefeller and his partners with complex legal problems. Ohio Standard's charter permitted the company to manufacture, transport, and market petroleum products in Ohio, but it could neither hold stock in other corporations nor conduct out-of-state business. As Standard Oil acquired firms outside Ohio, it had to place the stock of those companies in the hands of trustees, who then held that stock for the combination. But the arrangement was strictly informal, and there was no legal guarantee of a trustee's subordinance to the organization. As the number of out-of-state businesses in the hands of individual trustees mounted, the system became increasingly unsatisfactory. If a trustee died or decided to pursue an independent course, Standard Oil would be in trouble. Hence, in 1879, the Standard leadership placed all stock in the hands of three relatively minor company officials. The three trustees were mere figureheads whose sole function was to hold stock and distribute dividends.[3]

Meanwhile, other serious organizational problems plagued the combination. Standard companies around the United States and in several foreign countries retained considerable independence. Minority interests exercised undue influence in several subsidiaries. Standard marketing affiliates frequently invaded each other's territories and slashed prices in order to increase sales. And although Standard exported over 50 percent of its kerosene, the combination lacked a coherent program to develop foreign markets. Furthermore, Standard leaders wanted to transfer their headquarters from Cleveland to New York, a location closer to the main channels of international trade.[4]

S. C. T. Dodd, who joined the Standard legal staff in 1879, devised a plan to deal with these problems: the Standard Oil trust agreement. Signed on January 2, 1882, this document established a trust to hold the securities of forty-one investors in forty corporations. The stockholders received a total of $70 million in trust certificates for their securities. Since the various Standard companies had a net value of $55,221,738 in 1882, these certificates represented almost $15 million in watered stock. But the large earnings and rapidly mounting assets of the combination soon far exceeded the official $70 million capitalization. In the 1880s the Standard Oil trust was the largest and richest manufacturing organization in the world.[5]

The trust agreement of 1882 created a board of nine trustees to be elected by the certificate holders. John D. Rockefeller, William Rockefeller, Henry M. Flagler, and John D. Archbold were the most prominent members of the early board. They collected interest and dividends on the securities held in trust and determined the amount to be redistributed to the certificate holders. They also wrote the bylaws of the organization and filled vacancies on the board occurring between elections. Most important, the trust agreement granted them full managerial powers.[6] The board, from its new headquarters in New York, imposed centralized administration on the rapidly growing Standard empire for over a decade.

Like the combination itself, the antitrust campaign against Standard Oil began in Ohio. In October 1889 the state attorney general, David K. Watson, discovered the text of the Standard Oil trust agreement in the appendix of a legal treatise.[7] He immediately recognized that this agreement was legally vulnerable for two reasons: it was void and unenforceable at common law because it created a monopoly, and it forced Standard Oil of Ohio to violate its charter by operating under the direction of an out-of-state authority, the Standard Oil board of trustees. Watson immediately began work on a suit against Ohio Standard.[8]

The realities of Ohio politics offered the attorney general little encouragement as he prepared his case. The Republican party dominated the state, and big business dominated the party. The GOP leadership reflected this business orientation. Joseph B.

Foraker, a successful Cincinnati corporation lawyer, served as governor of the state between 1886 and 1890. Marcus A. Hanna, a retired Cleveland industrialist, was also a dominant figure in Ohio Republican politics. And William McKinley, a congressman from industrial northeastern Ohio and a high tariff advocate, was a rising force in the party. Ohio Democrats were just as devoted to the interests of big business. In 1890, a utility magnate who lived in New York, Calvin S. Brice, succeeded Henry B. Payne, who had close Standard Oil connections, to the United States Senate. John R. McLean, the wealthy owner of the Cincinnati *Enquirer,* was another power among Ohio Democrats.[9]

Despite the unfavorable political climate, Watson filed a quo warranto petition, which is used to prevent the exercise of powers not conferred by law, against Standard Oil of Ohio in the state supreme court on May 8, 1890. He noted that Ohio Standard's articles of incorporation designated Cleveland as the principal place of business and fixed the number of shares at thirty-five thousand. But through the trust agreement of 1882, the company had transferred all but seven shares to the trustees in New York, most of them nonresidents of Ohio. Since that time these trustees had exercised total control over Ohio Standard. Watson therefore charged that the company had violated its charter and thus had "forfeited its right to do business in the state."[10]

One historian pictured the attorney general who challenged Standard Oil as a "St. George" who "rode forth against the dragon."[11] Another called him a "party maverick" responding "to the fears of some farmers, small merchants, and the white-collar class that feared the giant trusts were becoming too big and powerful."[12] Although antitrust sentiment undoubtedly influenced Watson, he was not a visionary crusader. Watson served as a Republican attorney general between 1888 and 1892, he was the successful GOP candidate for a congressional seat in 1894, and he was defeated for reelection in 1896 by a reform coalition of Democrats and Populists. Watson's writings reveal his conservative political orientation. In 1899 he published *History of American Coinage,* in which he expressed passionate sound money convictions.[13] Eleven years later, he enthusiastically endorsed the conservative doctrine of liberty of contract in a two-volume work on constitutional law.

More to the point, he applauded the rise of the great corporations: "As business instrumentalities and agencies corporations are as permanent as the States themselves. No form of business is destined to endure longer than they, and the American people must recognize that they should not be abused, but should be encouraged."[14]

Watson apparently considered his suit against Standard Oil a matter of duty: "I was young then and I supposed that it was expected of a public officer to perform his duty . . . . There was a statute in Ohio which required an attorney general to bring suit against any corporation which he had reason to believe was violating the laws of the state." Watson then added, "I had no personal feeling against the Standard Oil Company, but I meant to enforce the law against it as I would against any other company which I believed to be violating the law."[15]

An important series of court decisions, based on the common law, invited a legal assault on Standard Oil. State courts had recently ruled against the cottonseed oil trust, the whiskey trust, the sugar trust, and several other large combinations, finding them guilty of either violating general common-law prohibitions against monopoly or directing subsidiaries to transfer control of their operations out of state in violation of their charter.[16] These were the same charges that Watson made against Standard Oil.

Although these decisions gave Watson a strong legal position, he still faced tremendous pressure from the business-dominated political hierarchy of the state. A later Ohio attorney general, Frank S. Monnett, charged that Watson was offered bribes to drop the suit on six different occasions.[17] And on November 21, 1890, Watson received a letter from Marcus Hanna, a close friend of John D. Rockefeller, advising him not to push Standard Oil too hard.

There is no greater mistake for a man in or out of public place to make than to assume that he owes any duty to the public or can in any manner advance his own position or interests by attacking the organization under which experience has taught business can best be done. From a party standpoint, interested in the success of the Republican party, and regarding you as in the line of political promotion, I must say that the identification of your office with litigation of this character is a great mistake.[18]

Watson replied that he remained convinced that the suit was justified and that he did not intend any general assault on "organized capital." Hanna made a final effort to dissuade the attorney general: "Politically it is a very sad mistake, and I am sure will not result in much personal glory for you." But Watson moved ahead.

Standard Oil assembled a distinguished team of defense attorneys. Joseph H. Choate, who argued the case in court, was one of the great lawyers of his day, particularly noted for his brilliant oral presentations. He had served as president of the Association of the Bar of the City of New York, the American Bar Association, the New York State Bar Association, and the Harvard Law School Association. He was also a leading figure in the social and political affairs of New York City. President McKinley capped Choate's illustrious career in 1898 by appointing him United States ambassador to Great Britain. The other Standard Oil attorneys were S. C. T. Dodd, legal architect of the trust agreement and head of the trust's legal department, and Virgil P. Kline, chief counsel for Standard Oil of Ohio.[20]

The Ohio Supreme Court heard oral arguments in this case on October 8, 1891. The court handed down its decision on March 2 the following year. The first major question in the case was whether Standard Oil of Ohio, in its corporate capacity, entered the trust agreement of 1882. Watson argued that virtually all directors and stockholders had signed the agreement and that all but seven shares of stock had been transferred, activity that "constituted actual corporate conduct, if not formal corporate action."[21] Choate responded that only individual stockholders acting in their individual capacity had entered the trust agreement. The corporation was a legal entity distinct from its stockholders, and only the acts of corporate agents affected the corporate entity.[22]

Judge Thaddeus A. Minshall, who wrote the unanimous decision, decided this point in favor of the state. He ruled that the concept of corporate entity was merely a useful legal fiction and was thus subservient to "the ends of justice." In this case justice required that the fiction be overlooked. "Is there room for doubt, that the act of all the stockholders, officers and directors of the company in signing [the trust agreement], should be imputed to them as an act done in their capacity as a corporation? We think not, since thereby all the property and business of the company is, and was intended to be, virtually transferred to the Standard Oil Trust, and is controlled, through its trustees, as effectually as if a formal transfer had been made by the directors of the company."[23] The stockholders' act therefore was a corporate act in violation of the charter.

The second question concerned the issue of monopoly. Watson charged that an agreement violated the common law and was thus void when it created "a substantial monopoly, or is in restraint of trade or otherwise injurious to the public."[24] He apparently considered Standard's total dominance of Ohio's oil industry beyond dispute because he presented no economic evidence to support his charge of monopoly. Similarly Choate did not try to deny Standard's monopolistic position. Still denying any corporate act, he argued that the trust agreement did not create a partnership between corporations; it was merely a financial arrangement between private individuals, stockholders and trustees.

Minshall again ruled in favor of the state. He asserted that the object of the trust agreement was to establish a "virtual monopoly" in oil refining, that the trust now successfully controlled the supply and price of oil, and that "all such associations are contrary to the policy of our state and void."[25] Minshall then offered general remarks, worthy of a good Jacksonian, on the question of monopoly:

It may be true that it [Standard Oil] has improved the quality and cheapened the cost of petroleum and its products to the consumer. But such is not one of the usual or general results of a monopoly, and it is the policy of the law to regard not what may, but what usually happens . . . . A society in which a few men are the employers and a great body are merely employees or servants is not the most desirable in a republic; and it should be as much the policy of the laws to multiply the numbers engaged in independent pursuits or in the profits of production, as to cheapen the price to the consumer.[26]

The third major issue in the case involved the statute of limitations. Ohio law required that a suit to revoke a corporate charter begin within five years of the illegal act. In addition, a suit could not be filed against a corporation for exercising a power under its charter that it had already exercised for a period of twenty years.[27] Choate shrewdly argued that under this provision the suit against Ohio Standard had to commence within five years of the date of the trust agreement—January 4, 1882. Watson replied by citing a case showing that the statute of limitations did not pertain to an ongoing violation of charter. He then added the inept argument that the state had no knowledge of the trust agreement until late 1889, "and such statute does not begin to run until the frauds are discovered."[28] Choate countered with the general rule "that where fraud or fraudulent concealment of cause of action is relied upon to stop the running of the statute, the fraud must be specifically averred."[29]

Minshall steered a middle course in resolving this tangle. He ignored the matter of ongoing violations. He agreed with Choate that the statute of limitations blocked forfeiture of charter and that the question of fraud was irrelevant because fraud had not been formally charged. But he also ruled that by entering the trust agreement, Ohio Standard had "exercised a power for which it had no authority under the laws of this state."[30] Ohio Standard had committed this act within the last twenty years. Therefore the court ordered the company to withdraw from the trust.

Although the Ohio Supreme Court failed to revoke Standard Oil of Ohio's charter, both contemporary observers and later historians agreed that the judges had taken a firm stand against the trust. Ida Tarbell, a passionate critic of the Rockefeller organization, wrote approvingly of the court's "broad assertion of public policy" concerning monopoly.[31] But such observations mistook form for substance. The court was following a well-established line of precedents in ruling for the state on the claim of corporate entity. For example, in *People* v. *North River Sugar Refining Company* (1890), the defense also had claimed that the corporation itself had not entered the sugar trust because only the shareholders signed the trust agreement, and the corporate entity was not responsible for the acts of shareholders. The New York Court of Appeals had rejected this argument, declaring that the defense asked it to rule that "while all that was human and could act had sinned, yet the impalpable entity had not acted at all and must go free."[32] And in numerous other cases, courts rejected the concept of a corporate entity distinct from the actions of its shareholders.[33]

The Ohio court, impassioned rhetoric aside, also followed the conventional path on the issue of monopoly. In the late nineteenth century, American judges generally believed that the common law had favored freedom of trade throughout the ages. Minshall thus expressed a common belief when he declared that "monopolies have always been regarded as contrary to the spirit and policy of the common law."[34] He was also able to cite the recent decisions regarding salt and candle combinations—and numerous precedents from other states—to prove that monopolies were contrary to public policy in Ohio and thus void.[35] Contemporary scholars surveying the American application of the common law on monopolies frequently pointed to the rigorous restraints the courts imposed upon combinations. Even S. C. T. Dodd recognized that the Ohio court was in general agreement with other American courts. In an 1894 article in the *Harvard Law Review,* he lamented the negative attitude of the American judiciary concerning monopolies and implored the judges to ease the traditional restraints of the common law.[36] But Minshall and most other American jurists were not ready to alter the law to accommodate the captains of industry.

The court displayed a singular lack of hostility toward the trust on the question of the statute of limitations. Choate seriously distorted the charge against Ohio Standard when he argued that entering the trust agreement of 1882 was his client's sole offense. A glance at Watson's petition reveals that this was not the only violation. The petition read: "said nine trustees have been ever since the signing of said agreements, *and still are,* able to choose and have chose, *annually,* such boards of directors of defendant company as they (said nine trustees) have seen fit, and are able, *and do,* control the action of the defendant in the conduct and management of its business."[37] Hence the state charged Ohio Standard both with entering the trust and continuing operations under the direction of the board of trustees. Watson noted Choate's distortion of his position by citing *State* v. *Railway Company,* a case in which an Ohio railroad had lost its charter because of an ongoing violation. Although proceedings against the railroad began more than five years after incorporation, the statute of limitations did not bar forfeiture.[38]

Judge Minshall, however, chose to ignore Watson's reference to *State* v. *Railway Company.* Had Minshall taken the firm stand against the trust usually attributed to him, he surely would have seized this opportunity to avoid the statute of limitations. He then could have annulled Ohio Standard's charter rather than simply ordering its withdrawal from the trust agreement. In reality, Minshall had no interest in battling the trust. He followed the precedents on the first two questions in the case and sought a middle course on the third. Thus the Ohio court took a more moderate stand than had judges in similar cases where charters were forfeited.[39] The court showed further leniency by refusing to place a time limit on Ohio Standard's withdrawal from the trust. In response to Standard attorney Kline's March 16 request for more time, Chief Justice William T. Spear wrote: "So long as those in control appear to be engaged, as now, in an honest effort to dissever the relations of the company with the trust, and liquidate and wind up the affairs of the trust, the court will not be disposed to interfere."[40]

This moderate attitude reflected the basic character of the Ohio court. The five judges who sat on the bench were part of the conservative political establishment. Scattered biographical references reveal a group of conscientious but unspectacular jurists who followed the generally conservative judicial trends of the 1890s.[41] During the decade following the Standard Oil case, this court overturned a progressive inheritance tax, a mechanic lien law, a provision for the use of voting machines, and an eight-hour law for public works. An American Federation of Labor survey revealed that through 1916, the Ohio court had declared 132 state laws unconstitutional, making it the ninth most active in the use of the judicial veto. And a delegate to the Ohio Constitutional Convention of 1912, surveying volumes 63 through 84 of the *Ohio State Reports,* found thirty-three cases in which a circuit court ruling in favor of an individual was reversed by the state supreme court in order to protect a corporate interest.[42]

The court's refusal to revoke the charter, and its subsequent failure to impose a time limit on its withdrawal order, provided the oil trust with an opportunity to adopt a policy of systematic noncompliance. On March 21, 1892, the trustees and certificate holders met at the headquarters of the combination at 26 Broadway, New York City. Because Ohio Standard was an integral part of the total organization, they voted unanimously to dissolve the trust. The board of trustees would supervise the distribution of property, thus remaining in office as liquidating trustees. After four months, however, the trustees would lose the power to vote the stock that they still held.

In March 1892 the trust included eighty-four separate companies. On April 1 a major consolidation took place. The liquidating trustees transferred the stock of sixty-four companies to Standard Oil of New Jersey, Standard of New York, and several other major affiliates, leaving the trust with only twenty component corporations. Lawyers for the federal government later concluded that through this action, the trust carried out "a scheme to evade the effect of the decision of the Supreme Court of Ohio while appearing to comply with it. The fact of the transfer of these 64 companies was evidently a part of that scheme in order to reduce the number of companies, which they undoubtedly intended thereafter to operate in harmony."[43]

At the time of the dissolution, there were 972,500 trust certificates valued at a hundred dollars each. The trustees therefore divided the stock of each of the twenty remaining companies into 972,500 parts. A certificate holder could exchange a trust certificate for a form of assignment that entitled him to 1/972500 of the stock in each of the companies held by the trust. He could not be assigned stock in a single company without receiving the same proportion of stock in each company. About sixteen hundred people held Standard Oil trust certificates, but a group of seventeen, including the board of trustees, members of their immediate families, and top management personnel, owned more than half of the total. John D. Rockefeller alone held 256,785 shares. By December 31, 1892, the seventeen major shareholders, whose large holdings enabled them to acquire full shares of the constituent companies, had liquidated 494,619 certificates. None of the other shareholders had acted. Four years later the situation remained unchanged.[44]

The trustees could have spurred action by withholding dividends on trust certificates. Instead they developed a system of dividend payments that discouraged liquidation. They did not pay dividends on fractional shares; an investor had to own 194.5 trust certificates to receive a full share in each of the twenty companies. Furthermore, the trustees managed the combination's finances in a way that produced exceedingly irregular dividends from the constituent companies, while holders of unliquidated trust certificates received regular 3 to 5 percent dividends every quarter. For example, on September 15, 1897, the reported dividends were: Buckeye Pipe Line, 40 percent; Northern Pipe Line, 23.25 percent; Eureka Pipe Line, 12 percent; Northern Ohio Natural Gas, 1.5 percent; the other companies paid no dividends at that time. These percentages yielded $2,389,033.35 on the unliquidated shares. The trustees paid the certificate holders $5 per share (5 percent), totaling $2,389,400.[45] This system discouraged the small certificate holders from liquidating, obtaining fractional shares, and then trading for complete dividend-yielding shares of a single company, trading that would have scattered the stock of the various companies and undermined the trustee blueprint for continuing combination. The system made no difference to large shareholders who owned an equal proportion of stock in all twenty companies.

This state of perpetual liquidation also allowed the trustees to control a voting majority of the stock in all twenty companies. They could no longer vote the stock still held in trust, but as a group they controlled over 50 percent of the stock of all twenty companies. After 1892 the trustees performed the same duties as before dissolution. The top executives all remained at the same headquarters, and the trust continued exactly as before. The testimony of John D. Archbold, a member of the Standard board of trustees, before the United States Industrial Commission in 1899 aptly characterized the condition of the organization.

Question: Nevertheless, since that time the different Standard Oil companies have worked together in harmony, have they not?

Archbold: The ownership has naturally brought them into harmony of action; the like ownership, of course.

Question: The general way in which the control has been kept uniform has been this, that the men who were the former trustees have held the majority of stock in each of one of these different companies?

Archbold: Exactly so.

Question: So that the Standard Oil combination, as we may say, has worked together as harmonically since the dissolution of the trust as before?

Archbold: It is hardly fair to call it a combination, but you might call it an aggregation.

Question: An aggregation?

Archbold: An aggregation.

Question: But as a matter of fact it has worked as harmoniously as before?

Archbold: Yes.[46]

George Rice, an independent oil refiner and a persistent antagonist of Standard Oil, was not impressed by the "dissolution" of 1892. He was determined to do everything in his power to guarantee the actual destruction of the combination. In October 1892 Rice obtained six shares in the Standard Oil trust. He did not know the details of the trust's reorganization, but possession of the shares gave him a chance to keep a wary eye on his powerful adversary. Despite the dissolution, the trustees did not request that Rice liquidate his holdings. Moreover, he received quarterly dividends. Rice then attempted to liquidate one share by having it split into partial shares of the twenty constituent companies. Standard officials strongly resisted, complaining of the small fractions involved in this transaction, but Rice had the Ohio Supreme Court decree on his side. In April 1896, he finally received his fractional shares. This chain of events convinced Rice that the trustees did not intend to liquidate. In late 1897 he consulted several attorneys, including David K. Watson, who advised him to lay the facts before Frank S. Monnett, Ohio's current attorney general.[47]

Monnett had been elected to office in 1895 on the Republican ticket. Unlike Watson, he was passionately committed to the antitrust crusade.[48] He was also aware of the political possibilities of the antitrust issue. National concern about industrial combination had peaked in 1890 with the passage of the Sherman Act, and the depression of 1893 had hastened the eclipse of the issue. But by 1897 economic conditions were improving. In the following six years a wave of business consolidation occurred that fundamentally altered the legal and financial structure of American industry. The nation's attention focused on the trusts as never before.[49] Although big business still dominated Ohio politics, Monnett knew that the tide of public opinion temporarily flowed in his direction. He meant to use the antitrust issue to advance his political career.[50]

Armed with ideological conviction, political ambition, and a firm sense that the Standard Oil trust had defied the law, Monnett met in chambers with the judges of the Ohio Supreme Court. He presented Rice's evidence and argued that it justified proceedings against Ohio Standard. The judges agreed, and Monnett filed a contempt suit on November 9, 1897, charging that Ohio Standard had never intended to withdraw from the trust. The liquidating trustees, under the guise of dissolution, merely had adopted a plan to prevent the destruction of the combination.[51]

Monnett simultaneously pursued the trusts on another front. He and a state senator, H. E. Valentine, pressed for an investigation of anticompetitive business practices. They succeeded, despite strong opposition, in having a senate investigating committee appointed. In January 1898 this committee examined seventy-eight witnesses, who gave testimony on the railroad, insurance, coal, and petroleum industries. The investigation, supplemented by information gathered in subsequent antitrust suits, produced a reasonably comprehensive picture of the oil trust's operations in Ohio. Four Standard companies operated in the state. The Ohio Oil Company contracted with well owners for crude oil. The Buckeye Pipe Line Company then transported the crude to refineries, and the Standard Oil Company of Ohio and the Solar Refining Company operated those refineries. These firms worked as a unit under the direction of the Standard board of trustees. Buckeye Pipe Line controlled over 85 percent of the pipelines in the state. Solar Refining and Ohio Standard owned about 80 percent of the refinery capacity. Because it purchased, transported, and refined most of Ohio's oil, the Rockefeller organization set the price of both crude and refined petroleum. Investigators found widespread evidence of price manipulation, transport rate discrimination, and exorbitant profits.[52]

The investigation resulted in the Valentine-Stewart Antitrust Act, which was passed on April 19 and took effect on July 1, 1898. The Ohio law defined a trust as a "combination of capital, skill or acts" that restrained commerce, limited production, manipulated or fixed prices, prevented competition in manufacturing, or contracted to limit sales or pool resources. Such trusts were "unlawful, against public policy and void." In the event of a violation, the attorney general was to initiate a quo warranto suit for forfeiture of charter. Violators were also subject to fines and imprisonment. In a direct slap at the Standard combination, the act declared it unlawful "to issue or own trust certificates" or to place the management of a combination "in the hands of any trustee or trustees with the intent to limit or fix the prices or lessen the production and sale of any article of commerce."[53]

With an antitrust law now on the books, Monnett again turned his attention to the contempt suit. In November 1897 the Ohio Supreme Court had ordered John D. Rockefeller, president of Ohio Standard, to respond in writing to a list of questions posed by the attorney general. In February 1898, Rockefeller reluctantly disclosed that during the five years preceding the filing of the contempt suit, only one of the 477,881 uncancelled Standard Oil Trust certificates had been turned in—the one George Rice had redeemed. Yet in the three months following that suit, over 100,000 certificates were exchanged for fractional shares.[54] This behavior indicated that the trust was more interested in avoiding legal trouble than in dissolution.

Rockefeller's written answers, however, revealed little else about the trust's operation. Monnett therefore requested that the court appoint a commissioner to take direct testimony from the oil baron in New York City, and the court directed Rockefeller to appear at the New Amsterdam Hotel on October 11, 1898. The "richest man in the country" was on the stand for two days. He was polite but not very informative, prompting the New York *World* headline:

"John D. Rockefeller Imitates a Clam." Monnett managed to obtain a list of the certificate holders who had exchanged shares after November 1897, but he got very little else.[55]

Back in Ohio, Monnett moved against the other Ohio-based Standard corporations. In November 1898 he filed quo warranto petitions in the state supreme court, as outlined in the Ohio antitrust act, against the Ohio Oil Company, the Buckeye Pipe Line Company, and the Solar Refining Company. In January 1899 he brought suit against Standard Oil of Ohio. The attorney general demanded forfeiture of their charters on several grounds: the companies were members of a trust and thus acted contrary to public policy; they operated in violation of the antitrust law; they had conspired to evade the court decree of 1892; and the Buckeye Pipe Line Company had acted beyond the provisions of its charter by engaging in the telegraph business as a common carrier.[56]

Monnett, supported by the feisty George Rice and Senator Valentine, had launched an imposing offensive against the oil combination. The tide of public opinion flowed temporarily in his direction. The Ohio court had agreed that contempt proceedings were in order and had appointed a commissioner to take testimony. But the oil trust was no stranger to the legal battlefield. In late 1898 Standard officials embarked on a counteroffensive that culminated in total legal victory and the permanent destruction of Monnett's political career.

The first step in the Standard campaign was to plug further leaks of damaging evidence. In New York Monnett had extracted from a reticent John D. Rockefeller the admission that 13,593 shares of Ohio Standard stock were still in the pre-1892 form of dividend-yielding trust certificates. Did Ohio Standard pay money directly to the board of trustees in New York to cover these dividend payments? Monnett pressed the investigation on this point. Ohio Standard officials claimed they had paid no dividends since 1892; trust certificate earnings came exclusively from the other nineteen companies in the combination.

On November 15, 1898, Monnett demanded that Ohio Standard prove this unlikely contention by producing its books for the years 1892 to 1897. F. B. Squire, the company secretary, refused. The attorney general wanted Squire immediately jailed for contempt, but the commissioner in charge of the investigation declined to take action. Monnett then appealed to the state supreme court, which on December 7 ordered Squire to hand over the books. Squire still refused. His attorney claimed that delivery of these records would violate Squire's right against self-incrimination and would constitute an unreasonable search and seizure. The frustrated attorney general could do nothing but file a contempt suit against Squire.[57]

On December 21, 1898, Monnett created a statewide sensation by charging that Ohio Standard had burned the books sought by the prosecution. He claimed that an "anonymous communication" had informed him that this willful destruction of evidence had taken place on November 19 and 21.[58] Monnett promptly took depositions on this matter from several witnesses in Cleveland. A Bohemian rabbi swore that he had met a teamster in a saloon who talked of taking company books to a furnace where they were burned. The saloonkeeper corroborated the rabbi's story. And John McNirney, a Standard employee, explained how he had removed and burned sixteen boxes, as well as assorted papers and letters. "It is a fair legal presumption from the testimony," Monnett concluded, "and from the size of the books and the size and number of the boxes, and the circumstance of their refusing to produce them afterwards" that important evidence had been destroyed.[59]

The Standard attorneys issued a prompt denial. Virgil P. Kline declared in the Cleveland *Press:* "The charge or insinuation made by Attorney General Monnett that the Standard Oil Company had burned or destroyed any of its books of account is absolutely false. It is simply based upon the fact that the company, from time to time, destroys useless material which accumulates in its business."[60] The company then produced all the workers who participated in the burning, and they swore that the operation was routine. The company even induced McNirney to testify that the boxes he had burned contained only "waste paper."

Although the attorney general had a strong circumstantial case, the truth of Monnett's book-burning charge is impossible to determine. The Standard combination certainly wanted to keep the records away from Monnett at all costs because the books would have proved all four Ohio-chartered Standard companies guilty of violations of the state antitrust law. And the testimony of Standard employees, who faced the possible loss of their jobs, does not go far to exonerate the company. Still, Monnett failed to produce conclusive evidence to substantiate his charge. By going to the press, armed with an "anonymous communication" and a few shaky witnesses, Monnett seriously damaged his public image. Ohio Standard's defense of its books was a complete success. Squire remained adamant, and the contempt suit against him offered little hope of producing immediate results.

Meanwhile the Standard organization bolstered its political alliances. In December 1898 John D. Archbold procured the legal talent and political influence of Joseph B. Foraker, who was then serving in the United States Senate. Foraker later explained that Ohio Standard merely had employed him to investigate the legal situation in the contempt and antitrust suits.[61] Although he frequently counseled the other Ohio Standard attorneys, he did not participate directly in the litigation. Foraker's service to the company continued until 1901, and his fees were substantial; Ohio Standard paid him $44,500 in 1900 alone.[62]

Foraker earned part of his salary by meeting with Monnett in Washington in January 1899. At this meeting, Foraker told Monnett of his new job with Ohio Standard and used his leverage as a leader of the Ohio Republican party and as a United States senator to induce Monnett to delay the contempt and antitrust suits. Monnett refused to cooperate, and Foraker threatened political reprisal. Monnett gave the following account of the meeting:

I at first discussed the impropriety and danger of his representing these trusts, criminal and civil violators of his own State, as long as he as well as myself should be interested in the welfare of the people of Ohio. He told me that he never allowed his law practice to interfere with politics or his politics with his law practice, and added that he was a judge of ethics of our profession. He then took up the cause of action against these companies and reminded me of the great power financially and politically of the Standard Oil crowd. After talking a short time he asked me to have the proceedings delayed in order to accommodate him. I firmly declined to concede any time whatever and told him so. He recalled the great power of the Oil Trust to anyone opposed to it.[63]

The Standard combination also took steps to guarantee that it received the support of the Ohio press. In early December 1898 over a hundred Ohio and Indiana papers featured news items and editorial comment favorable to Standard Oil and critical of Monnett. A representative item appeared in the Lima *Times Democrat:*

Whether the Standard Oil Company of Ohio is in a trust or out of a trust is a question for the courts to decide; but whether the consumers of oil are getting a better quality at less cost and handling with greater safety than formerly is a question for the people to decide. In the commercial affairs of life it is things, not words, that count in making up the balance sheet of loss or gain, of benefit or injury. Monopoly and octopus, combines and trusts, are haughty words, but the best goods at lower prices are beneficial things. It is much easier to say harsh words than it is to make good things cheap.[64]

Monnett discovered that this and similar articles were the product of a campaign conducted by the Malcolm Jennings Advertising Agency. Ohio Standard hired Jennings, who in turn contracted with newspapers to advertise paraffin and other company products. These contracts included provisions for space on news and editorial pages where statements favorable to Standard Oil would appear without identification as advertisements. Monnett acquired the agency's contract with the *Xenia-Herald:*

The publisher agrees to reprint, on news or editorial pages of said newspaper, such reading notices set in the body type of said paper and bearing no marks to indicate advertising, as are furnished from time to time by said Jenning's agency, at a rate of———per line, and to furnish such agency extra copies of paper containing such notices at 4 cents per copy, or to mail the same to a list of subscribers furnished by said agency at 4 cents per copy.[65]

The attorney general summoned Jennings to testify. Although the advertising agent admitted the existence of these contracts, he claimed the practice was common, and he insisted that editorial opinion had not been furnished. The provisions for printing his articles on the editorial page simply was to stop newspapers from placing them in the miscellaneous section. Monnett demanded a list of all the newspapers that the agency had dealt with. Jennings refused, claiming that such an order invaded his private rights.[66] Monnett then charged the advertising agent with contempt and threw him into jail, but Jennings was soon released on a writ of habeus corpus. In early May 1899, the Ohio Supreme Court dismissed the contempt charge, claiming that the attorney general's questions were irrelevant to the main issues in the case.[67] The Ohio Standard press campaign continued unimpaired.

In the midst of the Standard counteroffensive, Rice and Monnett made some mistakes that seriously damaged their cause. On March 3, 1899, Rice went to the press with the most sensational accusation of the whole Ohio controversy: "It might be of interest to the citizens of the State to know that they have an incorruptible official in Attorney-General F. S. Monnett, who within the past month has been offered the sum of $500,000, less $100,000, to be retained by the person attempting bribery, to stop proceedings against the Standard Oil interest."[68] Monnett was surprised and angered at this careless disclosure of confidential information.[69]

The story immediately received national press coverage. Reporters rushed to interview Monnett, and the representative of the New York *World* filed an account featured on the front page of the March 6, 1899, edition. According to the *World*, Monnett received a visit from a Cleveland "friend" at his office on the afternoon of January 20, 1899, who told him that certain individuals in New York had been commissioned by the Standard Oil Company to induce him to stop his suits. The Cleveland man had talked with the commissioners by telephone. They had offered Monnett $400,000 in cash, stock, or trust certificates. They would place the bribe in a safe-deposit box and send the key to Monnett. The attorney general declared that his brother-in-law, Smith W. Bennett, had witnessed the offer and was ready to back his story. But Monnett refused to divulge the name of the Cleveland party.[70]

The Ohio press greeted Monnett's story with a chorus of derision, doubtless partly the product of the Jennings Advertising Agency. But there was also a widespread, apparently unsolicited, call for the attorney general to reveal the identity of his mysterious Cleveland confidant. Monnett, recognizing the difficulty of proving his story, was reluctant to pursue the matter, yet he was stung by the ridicule of the press. "I can well understand," he exploded, "why newspapers that have sold out their editorial opinions for 10 cents a line cannot believe that anyone could withstand an offer to sell his opinions and principles for such a figure."[71] He decided to report the bribe to the Ohio Supreme Court.

With this controversy unresolved, Monnett traveled to New York for a final attempt to obtain information in the contempt suit. On March 17, 1899, he began an examination of John D. Archbold. But Archbold belligerently refused to give any information, and the proceedings quickly degenerated into a farce. The second day of testimony featured a heated exchange between Archbold and W. L. Flagg, Monnett's assistant. A reporter for the *World* appraised the fray:

He [Flagg] gave the oil magnate as good as he sent—or rather, as bad as he sent, "Low-lived cur" and "dirty, stinking liar," and other terms which men usually resent with their fists or other weapons were hurled back and forth for a quarter of an hour, during which time it would have surprised nobody if the enraged Standard Oil magnate and the lawyer whose passion he aroused had engaged in a rough and tumble fight.[72]

Monnett expected testimony from two independent oil refiners, but they refused to appear, fearing reprisals from the oil combination. The Standard lawyers seized this opportunity to challenge the attorney general to divulge the name of the man who disclosed the alleged bribe. Monnett again refused.[73]

On April 15 Monnett brought another avalanche of abuse upon himself by filing a bribery petition in the Ohio Supreme Court without identifying his now-famous Cleveland friend. The Sandusky *Register* sneered, "The attorney-general has filed the long-promised statement of the attempt to bribe him by the Standard Oil Company. It is as unsatisfactory as the play of Hamlet with Hamlet omitted. He gives no names, but says that certain parties or a certain party offered to do so and so."[74] Standard attorneys requested that the court require Monnett to name names. The attorney general then filed an amendment to his petition in which he identified Charles B. Squire, an intimate acquaintance, as the man who had approached him on January 20. Monnett named F. B. Squire, Frank Rockefeller, and Charles N. Haskell as the Standard agents who had offered the bribe. He then foolishly compounded his problem by charging, without conclusive proof, that David K. Watson had received numerous bribe offers to drop his common-law suit in 1892.[75]

Monnett's case disintegrated when Charles B. Squire issued a clarification of his role, denying that he had ever informed the attorney general of a definite bribe offer. Squire recalled that he had been approached by a "promoter of schemes" and had warned the attorney general to beware of him. He later learned that the promoter had no connection with Standard Oil. "He was merely 'fishing' in the hope of getting something for himself."[76] After Squire's clarification, the Ohio Supreme Court refused Monnett's request for an investigation. And, in December 1899, the court ordered Monnett's April 15 petition struck from the record, observing that there was no "competent evidence" to connect Standard Oil with a bribery attempt.[77]

Although the evidence on this matter is incomplete, some tentative conclusions are possible. Records of the chief counsel for Standard Oil, Mortimer F. Elliot, indicate that the trust's legal office knew nothing of the bribe attempt.[78] Hence direct involvement by the oil combination is unlikely. All accounts agree that Monnett and Charles Squire met on January 20, 1899, and discussed the issue of bribery in some fashion, and Monnett supported Charles Squire in the latter's denial of criminal involvement. Yet Monnett and Squire gave radically different accounts of the character of the bribe offer discussed at the January 20 meeting. As early as March 5 Monnett claimed that Squire had mentioned the involvement of several Standard "commissioners"—later identified as F.B. Squire, Frank Rockefeller, and Haskell—and a bribe of $400,000. Charles Squire came forward with the story of a single unnamed "promoter" and a vague offer of an unspecified bribe in late April, after Monnett had disclosed his identity. Although Monnett was prone to rash and irresponsible acts, he would not likely have used those names and that amount in the absence of any prompting. Thus Charles Squire probably supplied Monnett with the names and the amount on January 20. Later, when Squire discovered his "promoter" was a fraud, he quickly backed away from his earlier story, and Monnett was left in the sole possession of another unsubstantiated charge against the oil trust. Still, the attorney general was primarily the victim of his own blunders. He erred in revealing the bribe story to the rash George Rice; he erred again in backing Rice's revelations to the press; and he compounded all his mistakes by bringing up the attempts to bribe Watson. As in the book-burning controversy, Monnett's zeal to crush the trust seriously affected his judgment and damaged his credibility.

As Monnett's public image sank, Standard officials took a final, dramatic step to counter the wave of lawsuits. In June 1899 the board of trustees reorganized the Standard interests into a holding company.[79] The Ohio litigation was the immediate but not the only reason for the change. Since 1892 the Standard companies had maintained de facto unity because seventeen stockholders, many of whom served as liquidating trustees, held majority interest in all twenty constituent companies. But several of these oil barons were getting old, and there was no guarantee that their heirs would act in harmony. A holding company would ensure organizational cohesion in future years. It might also ease public hostility produced by the glacial pace of the trust's litigation.[80]

Favorable legal developments outside Ohio also prompted reorganization. In 1895 the United States Supreme Court, in *United States* v. *E. C. Knight Company,* ruled that production was not interstate commerce and was therefore beyond the reach of the Sherman Act.[81] Standard lawyers could now reasonably assume that a holding company engaged in the production and refining of petroleum products would not violate federal antitrust law. Furthermore, recent changes in New Jersey law created an ideal base for the new organization. Before 1888 state courts generally refused to allow a corporation to hold stock in other companies. But in that year the New Jersey legislature amended state law to permit holding companies. New Jersey also featured a nominal incorporation fee, as well as low, stable taxes. The state required only one company director to be a resident, and the publication of corporate annual reports was not obligatory.[82] Thus Standard Oil of New Jersey became the parent company in the new organizational structure.

In the late 1890s most industrial consolidations took the form of simple mergers. But reorganization as a holding company offered the Standard group several advantages. Mergers required majority approval by stockholders of all companies involved and presented the possibility of lawsuits by dissenters, while the formation of a holding company required only that Jersey Standard stockholders approve amendments to the company charter. Moreover, Jersey Standard's acquisition of stock control of the other nineteen companies was less complicated and less expensive than a complete merger. Acquisition of stock control could proceed over a period of years as opportunities arose, thus avoiding the hostile publicity that invariably accompanied a sudden direct fusion.[83]

On June 14, 1899, the directors of Jersey Standard, with the approval of the stockholders, raised the total stock of the company from $10 million to $110 million in the form of 1 million shares with a par value of $100 each. The directors authorized the exchange of 972,500 shares of this stock for outstanding trust certificates and fractional shares in the twenty constituent companies. The Jersey Standard stock had the same value—$100 per share—as the old trust certificates, and exchanges were made on a one-for-one basis. The transfers proceeded rapidly; 969,698 Jersey Standard shares were issued in the last six months of 1899. The directors also altered the Jersey Standard charter to allow them greater powers in order to run the new organization. The leadership remained intact. John D. Rockefeller became president of Standard Oil of New Jersey; John D. Archbold became vice-president. The other fourteen directors included six who had been leading figures in the combination since the 1870s.[84]

Monnett concluded his investigation in the spring of 1899 and prepared to present his case to the Ohio Supreme Court. The court that reviewed his evidence was substantially different from the one Watson had faced in 1892. Between 1892 and 1899, two judges left office and the membership of the court was increased, from five to six. The judges who filled the three vacant seats, Jacob F. Burket, John A. Shauck, and William Z. Davis, were all conservative Republicans. Shauck, who became one of the most influential members of the Ohio court, was particularly notable for his broad view of the proper scope of judicial review and his negative attitude toward social legislation.[85] The new judges brought a little more conservatism to the court, but the changes in membership was primarily important because the Ohio court now had only three members who had participated in the original 1892 decision. Thus half the court had no personal interest in the implementation of the original dissolution order.

A more significant change in personnel occurred at the Ohio Republican state convention in 1899. Frank Monnett failed to win renomination. The attorney general claimed that Foraker, making good his January 1899 threat, engineered his defeat. Monnett later told the New York *World*, "I was defeated for a renomination by Senator Foraker and his friends and only received 87 votes in the convention out of over 400."[86] There is no direct evidence to support this charge, but subsequent actions of Foraker and Standard Oil officials lend support to the attorney general's accusation. In 1901 Monnett joined the Democratic party and attempted to win its nomination for attorney general. Ohio Standard lawyer Virgil Kline, a Democrat, portrayed Monnett as an unprincipled opportunist. Kline even induced a Toledo friend to enter the same race in order to deprive Monnett of his Toledo support. Consequently Monnett's candidacy failed.[87] Standard Oil's hostility extended to Monnett's brother-in-law, Smith W. Bennett. In 1903, when Bennett sought the Republican nomination for attorney general, Archbold sent the following letter to Foraker:

We are surprised beyond measure to learn that Smith W. Bennett, brother-in-law of F. S. Monnett, recently Attorney-General of Ohio, is in the race for the Attorney Generalship of Ohio on the Republican ticket. Bennett was associated with Monnett in the case against us in Ohio, and I would like to tell you something of our experiences and impressions of the man gained in that case. If you know him at all I am sure you will agree that his candidacy ought not to be seriously considered from any point of view.[88]

Whatever the role of Standard Oil in Monnett's defeat, the important fact was that he left office on January 1, 1900. John M. Sheets, an orthodox Republican with a sympathetic view of the oil combination, was Monnett's successor.[89]

In the Standard Oil litigation, 1900 was the year of decision. On January 30 the Ohio Supreme Court handed down a ruling in the four antitrust suits. Although the decision was announced after Monnett left office, the court heard arguments in late 1899. Hence

Monnett appeared for the prosecution. There were several parallel causes of action in each of the antitrust suits, but the court heard arguments on only one charge at this time: the state's allegation that each of the four Standard companies chartered in Ohio—Ohio Standard, Buckeye Pipe Line, Solar Refining, and the Ohio Oil Company—had entered the trust agreement of 1892 in order to prevent competition and fix prices in violation of the Ohio antitrust law. Attorneys for the defendant companies responded by proclaiming the Ohio law unconstitutional. The state demurred, and the court ruled only on the constitutional question.[90]

The court, in a succinct, unanimous decision by Judge Shauck, upheld the constitutionality of the Ohio antitrust law. Shauck limited his decision to the provisions of the law applicable to this case, which prohibited contracts to restrain trade and raise prices. "That contracts like these are hurtful," Shauck observed, "has been held in more cases than it would be practicable to cite. They abound throughout nearly three centuries of the development and administration of the common law in England and America."[91] He noted that the Sherman Act, which contained substantially similar provisions, had been recently sustained by the United States Supreme Court.[92] He concluded that the precedents overwhelmingly supported the state.

The *Weekly Law Bulletin*, an Ohio legal journal, pronounced the decision "a decided victory for ex-Attorney General Monnett."[93] A New York *World* headline proclaimed, "Blow Dealt to the Standard Oil Company in Ohio."[94] But the victory was costly. The constitutionality of the Ohio antitrust law was never seriously in doubt. Judge Shauck was noted for his strong belief in constitutional restrictions on legislative acts, yet even he had to bow to the numerous precedents upholding antitrust laws. Standard attorneys lost little in this decision, and they gained what they needed most—more time.

After January 1, 1900, prosecution of the four antitrust suits became the responsibility of Attorney General Sheets. Monnett had gathered a great deal of damaging evidence in these suits. For example, the facts indicated that the Buckeye Pipe Line Company, one of the four defendant corporations, controlled crude oil prices in Ohio because it purchased approximately 90 percent of the entire supply. As an independent oil producer put it: "We are at the mercy of the pipe line; they can control the market. They can put the oil down to twenty-five cents or they can put it up to two dollars and we have to take it."[95] Such manipulation of prices was clearly in violation of the state antitrust law, which prohibited combinations from acting to "increase, or reduce the price of merchandise or any commodity."[96]

Yet Sheets refused to bring the antitrust suits to trial. On December 21, 1900, these cases were "dismissed by the attorney general at cost of the state and without record."[97] Sheets also terminated the contempt case against F. B. Squire.[98]

Just ten days earlier, on December 11, the Ohio Supreme Court had finally ruled on Monnett's contempt suit against Ohio Standard for failing to comply with the 1892 court order to withdraw from the trust. The court split three to three, and the tie vote meant victory for Ohio Standard. The court did not give a formal opinion on the case.[99] The three judges who voted to hold the company in contempt—Minshall, Williams, and Spear—had all participated in the original dissolution decision in 1892. All three judges who had been elected since 1892—Burket, Shauck, and Davis—voted for the company. The Cleveland *Plain Dealer* concisely appraised the effect of the decision: "Mr. Monnett retires defeated."[100]

This unbroken string of setbacks temporarily dampened the enthusiasm of Ohio officials for legal forays against the oil trust. But around the country public hostility against the Rockefeller organization continued to mount, and after 1904 several states filed antitrust suits against various Standard affiliates. By 1906 Ohio once again had a state official ready to confront the oil combination. In January of that year Attorney General Wade E. Ellis publicly announced that he would gather the facts and then move against the trust.[101] Anti-Standard feelings apparently ran particularly deep in Ohio in 1906 because two county attorneys beat Ellis into court with singularly ineffective attempts to expel the Standard organization from the state. On April 28, the Lucas County prosecutor filed suit against the four major Standard affiliates in Ohio—Ohio Standard, Buckeye Pipe Line, Solar Refining, and

Ohio Oil—in a state circuit court. He called for the ouster of these firms on grounds that they had conspired to decrease oil production and increase oil prices. The suit collapsed a year later when the circuit court ruled that the summons directed at Ohio Standard had been improperly served.[102]

On July 5, the Hancock County prosecutor filed criminal charges against John D. Rockefeller and other top officials of Ohio Standard in the county probate court. Armed with a warrant for Rockefeller's arrest, the Hancock County sheriff went to New York, declaring his intention to seize the oil baron at the pier the minute he disembarked from a European cruise. Rockefeller, however, deprived the sheriff of his moment in the national spotlight. He voluntarily appeared at the trial, which began on October 8. Despite a reported attempt to bribe one of the jurors, the jury returned a guilty verdict on October 19. But two months later, on December 24, the court of common pleas at Finlay overturned the conviction on grounds that a probate court had no jurisdiction to try an antitrust case.[103] Before the court handed down this ruling, the Hancock County prosecutor had lashed out once again at the oil trust. The county grand jury returned an indictment against Rockefeller and the Ohio Standard management on November 14. The prosecutor's second effort, however, proved no more effective than his first. There is no evidence that this case ever came to trial.[104]

Attorney General Ellis began his attempt to drive out the oil trust in November 1906. He filed quo warranto petitions against Buckeye Pipe Line, Solar Refining, and Ohio Oil, seeking to prevent these companies from allowing Jersey Standard to continue to own their stock.[105] Ellis was a firm believer in strict antitrust enforcement, and he considered the Standard organization a particularly offensive monopoly. The attorney general optimistically reported that these suits would "effectively separate the Ohio companies from the trust."[106] Ellis argued the cases before the circuit court of Allen County on November 21, 1908. Instead of handing down an opinion, the court ruled, on February 8, 1909, that the attorney general should amend the petitions to include Jersey Standard as a defendant in each suit.[107]

Ellis's term expired at the end of 1908, and the Standard Oil suits became the responsibility of his successor, Ulysses G. Denman, a conservative Republican and a corporation lawyer who lacked Ellis's passion for antitrust prosecution.[108] Looking for a reason not to pursue these Standard Oil cases, Denman claimed that his office was incapable of serving Jersey Standard:

> The Standard Oil Company of New Jersey has never qualified to do business in Ohio, nor were we able, after a most thorough search, to find any person who claimed to be or who was reputed to be an agent of this company within the state. We, therefore, were at a loss to know how service might be made upon the company.[109]

The attorney general found this puzzle insoluble for five months. Then, in May 1909, the circuit court offered the reasonable suggestion that the bewildered process servers might find a Jersey Standard representative at a stockholders' meeting of any of the defendant corporations—Jersey Standard, after all, exercised total stock control over each company. But for the next three years Denman continued to claim that he could not find a single Jersey Standard agent in Ohio, and the suits never came to trial."[110]

Denman dismissed all charges against the Standard companies in 1911.[111] By that time the federal government was deeply involved in antitrust activity against the oil trust, and the entire matter was before the United States Supreme Court. Denman had a plausible excuse for dropping this suit because it merely sought to end Jersey Standard's control over the three Ohio companies, a task theoretically now being assumed by the federal judiciary. But the Ohio attorney general's professed inability to serve a summons on an agent of one of America's largest corporations over a three-year period demonstrates that he was more interested in avoiding court than confronting Standard Oil.

The contemporary press coverage of this litigation makes fascinating reading. The public was intrigued by the oil trust, and the legal controversies made headlines in both the Ohio press and leading national papers, like the New York *World* and the *New York Times*. But the coverage tended to focus on dramatic acts and statements. The beginning of a new suit, a bribery charge, or a passionate antimonopoly pronouncement by the Ohio Supreme Court received extensive coverage. Similarly, the spectacle of an upstart state attorney general berating John D. Rockefeller at an antitrust hearing made great copy. But the process that actually determined the outcome of these cases was far less dramatic. These disputes routinely dragged on over a period of years and were habitually determined on obscure, if not absurd, legal points. Thus the public record of these controversies often began on the front page of the *New York Times* and ended in the index of the *Annual Report* of the Ohio attorney general's office. And even a particularly alert newspaper reader of the period could have reasonably concluded that the oil trust was being unfairly victimized by a succession of overzealous Ohio state officials.

(2)

Trust Busting in Texas

The 1890s were years of hardship in the West. Low farm prices had plagued the region for years, and the serious depression of 1893 compounded economic problems. Like many other Americans, westerners increasingly blamed the trusts for their plight. Standard Oil received its full share of criticism. In this climate, Standard officials were probably not surprised when, in the mid-1890s, Texas brought a series of antitrust suits against Standard's St. Louis-based marketing affiliate, the Waters-Pierce Oil Company. Although it eventually became a great oil-producing state, Texas imported virtually all of its petroleum from Appalachia and the Midwest before the turn of the century. The Waters-Pierce Oil Company, under the direction of Henry Clay Pierce, received petroleum products from Standard refineries at Whiting, Indiana, and Cleveland and distributed them throughout the Southwest. By 1890 the company had captured over 90 percent of the Texas market for Standard Oil.

Henry Clay Pierce rose to a position of power in the petroleum industry at an early age. Born in New York in 1849, he moved to St. Louis when he was sixteen and found employment as a bank clerk. In 1867 he entered the oil business as a distributor for his father-in-law, who owned the first oil refinery west of the Mississippi. Two years later Pierce bought the company and conducted business under the name of H. C. Pierce & Company. His distributing trade increased so rapidly that in 1873 he formed a partnership with William H. Waters, a St. Louis businessman willing to supply desperately needed expansion capital. Although a half-owner,

Waters did not actively participate in the management of the firm. By the late 1870s, Waters, Pierce & Company marketed petroleum products in Illinois, Missouri, Arkansas, Oklahoma Territory, Louisiana, Texas, and parts of Mexico.[1]

The Waters, Pierce marketing region contained a widely scattered population inadequately served by the railroads. Yet Pierce recognized that the fast-growing Southwest had great sales potential despite the formidable transportation problems and took steps to capture and hold the entire area. He prevented rivals from gaining access to his market by expanding as rapidly as potential sales appeared. He traveled frequently throughout the Southwest promoting business and appointing local Waters, Pierce agents. Pierce also improved distribution methods. He was the first marketer to switch from wooden to iron oil barrels. He overcame the lack of rail transport by inventing, manufacturing, and utilizing horse-drawn tank wagons for the distribution of oil. Waters, Pierce & Company offered consumers a full line of refined petroleum products, including kerosene, lubricants, and greases. Pierce's energy and ingenuity soon made his firm the dominant oil marketer in the Southwest.[2]

Although Waters, Pierce purchased increasing quantities of oil from Standard refineries throughout the 1870s, it was not part of the Standard combination. In 1875 Chess, Carly & Company, a Standard affiliate based in Louisville, sent agents into Missouri to attempt to undersell Waters, Pierce. Henry Clay Pierce responded with an aggressive marketing campaign in Kentucky, and the Chess, Carly management soon requested a truce. In 1876 the leaders of the two companies agreed to sell their products on opposite sides of the Mississippi River. John D. Rockefeller, the guiding force behind Chess, Carly, carefully observed this marketing battle and was impressed by Pierce's abilities. Thus in 1878, when Pierce once again needed expansion capital, representatives of Chess, Carly & Company, as well as H. A. Hutchins and W. P. Thompson (two Standard agents from Cleveland), stepped forward with offers to purchase Waters, Pierce stock.[3]

In 1878 Pierce incorporated a new firm in St. Louis, the Waters-Pierce Oil Company, with a capital stock of $100,000. Pierce and Waters owned 40 percent of the stock; Hutchins and Thompson controlled 40 percent; Chess, Carly & Company controlled 20 percent. The managers of Waters-Pierce and Chess, Carly formalized their earlier division of marketing territory, using a map to trace the precise boundary. In 1882 the capitalization of Waters-Pierce rose to $400,000, while the same stock division remained in effect. That same year, after the formation of the Standard Oil Trust in New York, Hutchins, Thompson, and the stockholders of Chess, Carly transferred their holdings to the Standard Oil board of trustees. The next major change in the ownership of the company came in 1890 when William H. Waters left the firm. Pierce took some of Waters's stock, raising his own share in the company to 32 percent. The Standard combination acquired Waters's remaining shares. Despite the fact that the Standard board of trustees now controlled 68 percent of the stock, Henry Clay Pierce still managed the company in complete freedom.[4]

Pierce maintained his domination of petroleum marketing in the Southwest through brutal methods. His agents carefully monitored the shipments of independent oil merchants in order to determine his competitors' marketing areas. He then drastically cut prices in those areas in order to drive his opponents out of business.[5] W. W. Roberts, a Texas oil dealer, explained this technique: "The Pierce Oil Co. do their [*sic*] best to keep other oil out, and when they find other men offering to sell they will cut below them until he is out of their way, and then put the price up again."[6] Waters-Pierce agents also made sales contracts that granted rebates to merchants who agreed to carry only Standard products. In addition, Henry Clay Pierce bought the Eagle Refining Company but continued to run that firm as though it were a competitor. These methods were so successful that Waters-Pierce sold between 90 percent and 98 percent of the petroleum in the Southwest throughout the 1880s and 1890s, and the profits were substantial. With a capitalization of $400,000, Waters-Pierce earned $275,000 in 1883 and $338,000 in 1885.[7]

Pierce's tactics soon produced complications in Texas. In that state any company controlling 90 percent of its market faced a decidely hostile political climate. Venerable common-law prohibitions against monopoly long had been an integral part of state law.

The Republic of Texas had inserted an antimonopoly declaration in its constitution in 1836, and the state constitutions adopted in 1845, 1866, 1868, and 1876 all contained similar provisions.[8] In 1899 a proud Texan extravagantly praised the antimonopoly heritage of his state at the Chicago Conference on Trusts:

In the Constitution under which we in Texas live—handed down to us by the heroes of the Alamo and San Jacinto—we are taught that "monopolies are contrary to the genius of a free government, and shall never be allowed"; and we adhere with unhesitating loyalty to both the letter and the spirit of that declaration.[9]

Texas legislators played a prominent role in Washington in the 1888 drive for a federal antitrust law. Congressmen Jo Abbott and David Culberson each initiated an antitrust bill in the House of Representatives, and John H. Reagan introduced an antitrust measure in the Senate. In the Texas state elections of that same year, the Farmers' Alliance demanded control of trusts. And when the Texas legislature convened in January 1889, four different antitrust bills were introduced in the Texas House. The legislators quickly passed a composite draft on February 14, 1889, with only one dissenting vote. The Texas Senate passed a similar measure on March 22. James S. Hogg, the 300-pound attorney general of Texas, then took the lead in forming the final statute, which won approval on March 30, 1889. Texas thus became the second state in the nation to legislate against monopoly. Only Kansas, which had passed an antitrust law on March 2, 1889, acted sooner.[10]

The Texas law defined a trust as a combination that restrained trade, manipulated prices, prevented competition, or entered into anticompetitive contracts. In the event of a violation, the state attorney general or the district attorney was to initiate quo warranto proceedings for forfeiture of charter. A violation carried a penalty of $200 to $5,000, and each day of illegal activity constituted a separate offense. Other provisions banned foreign trusts from the state and declared void all contracts in restraint of trade. A person associated with a trust who "knowingly" carried out a prohibited act was subject to a fine of $50 to $5,000 and imprisonment for one to ten years. Agricultural producers were specifically exempted from the provisions of the bill.[11]

Jim Hogg received wide acclaim for his substantial contributions to the writing of the antitrust act. George Rice, the untiring critic of Standard Oil, sent his congratulations: "You deserve great honor and credit for having so successfully freed your state from the evils of monopoly. In the name of all the independent refiners of oil, I thank you."[12] The Democratic party rewarded Hogg by nominating him for governor in 1890. He defeated his Republican opponent in the November election by a margin of 262,000 votes to 72,000 and won reelection in 1892.[13]

Through his contacts with George Rice, Governor Hogg was fully aware of the Waters-Pierce Oil Company's business practices, and on November 21, 1894, Texas moved against the oil trust. Attorney General Charles A. Culberson obtained criminal indictments in the Texas district court at Waco against John D. Rockefeller, Henry M. Flagler, William Rockefeller, and other members of the Standard board of trustees. Also named were Henry Clay Pierce and five Texas division agents of Waters-Pierce. The state charged that the defendants had conspired to control the production and sale of oil, raise prices to artificially high levels, secretly operate ostensibly competing concerns, grant rebates, and temporarily slash prices to drive competitors out of business.[14]

Governor Hogg requested the extradition of John D. Rockefeller from New York and Henry M. Flagler from Florida, but the governors of New York and Florida both properly refused Hogg's request because neither Rockefeller nor Flagler had fled Texas. Hogg and Culberson, both lawyers, should have known that the leaders of the oil trust were beyond their reach. American courts have consistently refused to consider "constructive presence" as grounds for extradition; they uniformly required actual physical presence within the state requesting extradition at the time of the alleged crime.[15] Hence, charging the out-of-state leadership of Standard Oil with criminal conspiracy was an inept legal tactic. Not even Henry Clay Pierce could be reached because he conducted his business from St. Louis. After this initial extradition attempt failed, the state should have dropped the indictments because only the minor Waters-Pierce officials based in Texas could be tried. Fines and jail terms for local oil salesmen would do nothing to free the state from oil trust domination. Nevertheless, the state forged ahead with the case.

George Clark and John D. Johnson, who served as defense lawyers for the Waters-Pierce agents, gladly accepted this ill-conceived legal challenge because they were convinced that the Texas antitrust law was unconstitutional. State antitrust laws were untested in court at that time, and the Texas law contained such broad language that S. C. T. Dodd, Standard Oil's top lawyer, believed it unconstitutionally banned all large-scale business operations. The general legal climate of the 1890s heightened the defense attorney's expectations. The courts had taken an increasingly conservative stand during the decade, overturning many statutes that conflicted with the dominant laissez-faire ideology of big business. Moreover, the nature of the state's case allowed the defense to test the antitrust law without risking anything more than penalties for minor company officials.[16]

Clark and Johnson first moved that the case of E. T. Hathaway, one of the five Waters-Pierce Texas division agents, be severed from that of the other agents. The district court at Waco granted this request and scheduled Hathaway's jury trial for December 2, 1895. The Standard attorneys then used this case to attack the constitutionality of the Texas antitrust law. They argued that the law infringed on the congressional power to regulate interstate commerce, interfered with a businessman's liberty of contract, and, by exempting farmers, violated the equal protection clause of the Fourteenth Amendment. Clark and Johnson, who represented the company throughout the 1890s, would repeat these arguments many times as Texas pursued Waters-Pierce, and Waters-Pierce pursued Texas, through the courts. But the defense team had an inauspicious debut. The court disallowed their constitutional objections, and the jury convicted Hathaway. The judge acknowledged Hathaway's minor role in the conspiracy by fining him only fifty dollars. Nevertheless, Hathaway, on the advice of his attorneys, refused to pay the fine and went to jail.[17]

The oil attorneys then took Hathaway's case to the Texas Court of Criminal Appeals. The defense again heavily stressed the unconstitutionality of the antitrust law, but it also noted defects in the state's indictment. Under the antitrust law Hathaway, as an agent of Waters-Pierce, had to be charged with "knowingly" engaging in anticompetitive business practices. Yet the indictment failed to charge him specifically with a knowing violation; it simply declared that Standard Oil and Waters-Pierce had violated the law and that Hathaway was a Waters-Pierce agent. The court of criminal appeals found this technicality significant. On June 26, 1896, it reversed the conviction and remanded the case to the district court at Waco for retrial.[18]

In 1894 Attorney General Charles A. Culberson, whose office had drafted the defective indictments, became governor of Texas. Because the new governor had little interest in either Hathaway or the other four Waters-Pierce agents, the state made no attempt to pursue the case. Languishing in jail pending retrial, Hathaway secured a writ of habeus corpus from the United States District Court on November 24, 1896. A hearing was set for December 7, but the state, having had enough of this matter, dropped all charges against Hathaway on the day of the hearing. Inexplicably Texas officials allowed the charges against William Grice and the other Waters-Pierce agents to stand, and the Standard attorneys promptly seized the offensive. Grice, who had been free on bail pending trial, placed himself in the Waco jail and petitioned for a writ of habeus corpus.[19]

The United States District Court granted the writ and handed down a decision on Grice's case on February 22, 1897. Clark and Johnson at last achieved a victory on the constitutional question. Judge Charles Swayne ruled that the Texas antitrust law was void because its restrictions on the right to contract violated the due process clause of the Fourteenth Amendment. He also declared that the law violated the amendment's equal protection clause by exempting farmers from its provisions.[20]

But the Waters-Pierce triumph was short-lived. On February 21, 1898, the United States Supreme Court reversed the federal district court decision. Justice Rufus W. Peckham, who wrote the unanimous decision, strongly defended the independence of state courts, arguing that a federal court should grant habeus corpus only in exceptional circumstances. Grice's voluntary return to jail convinced Peckham that this "case is clearly nothing but an attempt to obtain the interference of a court of the United States when no extraordinary or peculiar circumstances exist in favor of such interference."[21] Peckham therefore ruled that the district court lacked proper jurisdiction to intervene. Furthermore, he declined to consider the constitutionality of the Texas antitrust law. Grice returned to the custody of the state, but no further action was taken against him.

The Culberson administration emerged from this slightly ludicrous series of decisions in a more favorable position than it had any right to expect. Although it had failed to obtain convictions, the Supreme Court had salvaged the state antitrust law.

Meanwhile, Texas had begun another more important suit against Waters-Pierce. On May 12, 1897, Attorney General Martin M. Crane filed a quo warranto petition against the company, seeking to revoke its permit to do business in Texas. The petition, filed in the district court at Austin, charged Waters-Pierce with four major violations of the antitrust law: it had entered the Standard Oil Trust agreement of 1882; it owned and operated the Eagle Refining Company and other bogus competitors; it made exclusive contracts with merchants who handled its goods; and it granted rebates to merchants who had entered those exclusive sales contracts.[22]

Judge R. E. Brooks, who presided at the jury trial held in early 1898, narrowed the case substantially in his charge to the jury. He ruled that the state had not produced sufficient evidence to prove that the company had joined the oil trust. He also noted that Waters-Pierce's acquisition of the Eagle Refining Company and other competing concerns did not violate any specific provisions of the antitrust law. Brooks therefore instructed the jury to base its verdict on the exclusive sales contracts and rebates. The jury considered the state's evidence conclusive on those points and found the company guilty. Waters-Pierce lost its business permit, the court "perpetually enjoining said company and its agents from doing business in Texas," but it continued to operate within the state pending appeal.[23]

In March 1898 Clark and Johnson repeated their now familiar Fourteenth Amendment arguments before the Texas Court of Civil Appeals, but Judge H. C. Fisher was unimpressed. He sustained the revocation of the Waters-Pierce business permit and upheld the Texas antitrust law as a legitimate exercise of the state police power. Fisher indulged in some of the antimonopoly rhetoric so fashionable in Texas, speaking in somber tones of the "unbridled license" of the industrialists. He declared that the Texas antitrust act was a proper response to the trust's misdeeds, which had "assumed such proportions" that they endangered "the welfare of the people."[24] The Waters-Pierce Oil Company met final defeat in the Texas courts when the state supreme court refused to grant the company a writ of error.[25]

Clark and Johnson, assisted by S. C. T. Dodd from Standard Oil headquarters in New York, presented their final appeal, again based on the Fourteenth Amendment, before the United States Supreme Court in January 1900. On March 19, 1900, the Supreme Court sustained the ruling of the Texas Court of Civil Appeals. Justice Joseph McKenna avoided any consideration of the constitutionality of the Texas antitrust law, declaring, "We are not called upon to answer these arguments or to condemn or vindicate the statutes on this record."[26] He sustained Texas's revocation of the Waters-Pierce business permit because "the right of a foreign corporation to engage in business within a state other than that of its creation depends solely upon the will of such other state."[27] Texas, which had granted the Waters-Pierce permit in the first place, could certainly revoke that right for violations of the antitrust act.

Texas seemingly had triumphed. The state antitrust law had survived repeated challenges, and the Waters-Pierce stronghold on the Texas oil business apparently had been broken. Some newspapers interpreted the decision as paving the way for greater state control over trusts throughout the nation. The Chicago *Tribune* declared: "A foreign corporation must obey the laws of the State in which it does business. If it does not it can be expelled. To that extent at least the States have power over trusts."[28] The New Orleans *Times-Democrat,* more optimistic than accurate, exclaimed: "Any State which chooses to pass a law declaring combinations that are 'in restraint of trade' to be illegal, can straightaway rid themselves of the combinations by appealing to its own laws."[29]

But Henry Clary Pierce was not willing to submit to the Supreme Court decision without a fight. His company had earned over $5 million in the 1895-1900 period, and profits of that magnitude were well worth defending vigorously. In St. Louis Pierce discussed his problem with David R. Francis, a wealthy Democratic politician who had served as the governor of Missouri. According to Francis, Pierce complained that he was having legal trouble in Texas and wanted the name of "some Texas lawyer" who could help.[30] Francis suggested Texas Congressman Joseph W. Bailey, and Pierce perceived the possibilities of such an arrangement. Although Pierce claimed that he needed a "lawyer," what he actually wanted was an agent with political influence in Texas. After all, Clark and Johnson had already taken the case to the Supreme Court—and lost. Only political intervention in Austin could help Pierce now. On April 23, 1900, Francis telegraphed Bailey, requesting that he meet Pierce in St. Louis at his earliest convenience.[31]

Joe Bailey had been first elected to the U.S. House of Representatives in 1890 on the same wave of agrarian discontent that had swept Hogg into the Texas governor's mansion. He continued to serve in the House throughout the 1890s, enthusiastically championing free silver and bankruptcy legislation for farmers and small businessmen. He opposed the high Dingley tariff because it would decrease imports, thus promoting the growth of the trusts. He opposed the annexation of Cuba and the Philippines partly because monopolies such as the American sugar trust would benefit. Bailey became House minority leader in 1897 at the youthful age of thirty-three, and by 1900 he had emerged as Texas's leading politician.[32]

David Francis had known Bailey for nearly a decade when he recommended the Texan to Pierce. They had met in Washington in 1891, and from the start Francis had been highly impressed by Bailey's political skills. He knew that the Texas legislator would be a valuable ally for Henry Clay Pierce. He also knew that Bailey needed money. In 1899 Francis had helped Bailey purchase the six-thousand-acre Grapevine Ranch near fast-growing Dallas. This property was an excellent long-term investment, but it put Bailey $96,000 in debt. The Texas lawmaker needed $3,300 to meet current obligations.[33]

On April 25, 1900, Bailey conferred with Pierce in St. Louis. The congressman, whose testimony before the Texas legislature is the only account of the meeting, listened to Pierce's version of his company's ouster and responded that "the people of Texas will not, and in my opinion, the people of Texas ought not tolerate the methods of the Standard Oil Company."[34] Pierce then claimed that the Waters-Pierce Company was completely independent of the trust. He showed Bailey the charge to the jury in the district court trial, in which Judge Brooks had withdrawn the allegation that Waters-Pierce was a member of the Standard combination. Pierce emphasized that the courts had convicted his company only because it had granted illegal rebates and had made exclusive sales contracts. Bailey was impressed by Pierce's argument. Quickly adopting a sympathetic attitude, he observed, "Our people do not want to drive any legitimate business out of the state."[35]

Bailey's flexibility was remarkable. Although Judge Brooks had withdrawn the state's trust membership charge, his action merely demonstrated that prosecution had failed to produce adequate evidence on that issue. It proved nothing about the ownership of the company. If Bailey had looked at the recent Supreme Court decision in the Waters-Pierce case, he would have observed that S. C. T. Dodd, counsel-in-chief for the oil trust, had presented a brief on behalf of Pierce's company. If Bailey had examined the Standard Oil trust agreement of 1882, he would have noticed that Waters-Pierce was in fact included in the compact. While these documents did not reveal that the Standard Oil board of trustees held a controlling 68 percent of Waters-Pierce stock, they certainly offered ample reason to be skeptical of Pierce's story. Moreover, even if Waters-Pierce was independent, the company still had been recently convicted of illegal acts, and it controlled over 90 percent of the Texas oil trade. Clearly, Bailey had no legitimate reason to welcome Waters-Pierce back into the state.

Delighted with Bailey's sympathetic reaction, Pierce asked the congressman to serve as his attorney. Bailey declined: "I practice law and do not practice influence."[36] But he offered to talk to some friends in Austin about the matter. The Texan then explained his financial problems and arranged for a loan totaling $5,000. He immediately took $3,300 from Pierce and made provisions to get the rest later. Bailey promised to meet Pierce in Austin within the week and discuss his forthcoming conferences with Texas officials.[37]

In 1900 Joseph D. Sayers was governor of Texas, and "Honest Tom" Smith was his attorney general. Both politicians loudly professed their devotion to trust busting. The governor wrote magazine articles deploring the evils of monopoly and organized a conference in 1899 at St. Louis to promote corporate regulation. The attorney general appeared before the Chicago Conference on Trusts in 1900 advocating stronger laws against the combinations. Down in Texas, Smith proudly noted, the state legislature had just strengthened its antitrust statute.[38] Still, ambitious politicians in a corporate economy could not afford to take their own antimonopoly rhetoric too seriously. While publicly denouncing the trusts, the Sayers administration treated Waters-Pierce with special care. When the company's original business permit had expired in June 1899, the attorney general could not grant a renewal because of pending antitrust decisions, so Smith issued an informal extension for the duration of the litigation. On March 19, 1900, the United States Supreme Court finally upheld the Waters-Pierce explusion from Texas. Yet Tom Smith was not impressed. Instead of ousting the company, he granted another extension until May 5, 1900.[39]

Joe Bailey thus had good reason to expect a sympathetic hearing when he discussed the Waters-Pierce issue with Sayers and Smith on May 1, 1900. The governor readily agreed with Bailey that this "legitimate and useful business" should not be driven from the state. The attorney general, who was actually responsible for business permits, was somewhat more reticent. Bailey first asked Smith to readmit the company on the sole condition that Pierce promise to obey the law. But the attorney general showed Bailey the judgment against Waters-Pierce that declared the firm "perpetually enjoined" from doing business in the state, and Bailey had to agree that Waters-Pierce in its present form could not be readmitted.[40] Nevertheless, he came away from the meeting convinced that Smith would allow a reorganized company to enter the state. Bailey thus advised Pierce that "the only lawful course open to you is to dissolve this offending corporation, organize you a new one, come into this State with clean hands and obey our laws."[41]

Pierce reorganized his company with impressive speed. Missouri law, under which Waters-Pierce was incorporated, allowed a solvent corporation to dissolve through the unanimous consent of the stockholders. Since the Standard Oil board of trustees owned 68 percent of Waters-Pierce stock, Pierce went to New York and obtained the executive committee's approval of his dissolution plan.

He then returned to St. Louis, and on May 17, 1900, voted all 4,000 shares of Waters-Pierce stock in favor of dissolution.[42]

Two days later a new Waters-Pierce Oil Company was incorporated, capitalized at $400,000. The company issued 4,000 shares of stock with a par value of $100 each—exactly the same number and value as the stock of the previous company. Pierce then purchased all 4,000 shares, thus creating the impression that he had bought out Standard Oil. Actually, John D. Johnson and S. C. T. Dodd made arrangements at the time of the reorganization to return the Waters-Pierce shares to the trust secretly. A few months later, on September 4, the transfer occurred. Keeping the shares in his name, Pierce secretly assigned 2,748 shares to Charles Pratt, a member of the Standard Oil board of trustees.[43]

The Standard combination capped its new arrangement with Waters-Pierce by placing R. H. McNall on the payroll of the reorganized Missouri corporation. McNall previously had served as clerk for Standard Oil treasurer H. M. Tilford, but immediately after the reorganization he became a "commercial agent" for Waters-Pierce. His business address was 75 New Street, New York City—the back entrance of the Standard Oil building at 26 Broadway. McNall received daily reports of Waters-Pierce sales and receipts, periodically received complete financial statements from the company, and handled correspondence between Henry Clay Pierce and the Standard board of trustees.[44]

The daily operations of Waters-Pierce were not affected by the reorganization, still Pierce was hopeful that his scheme would succeed. The attitude of Attorney General Smith offered considerable grounds for optimism. On May 15 Smith had granted the company yet another informal extension of its business permit—this time until May 31.[45]

Pierce, Johnson, and Clark met with Texas Secretary of State Dermott H. Hardy in Austin on May 31. They presented to him certified copies of the Waters-Pierce article of dissolution and the new articles of incorporation. They also submitted an application for a new business permit. Hardy took the documents to the attorney general, and he concluded that there was no legal way that Texas could withhold a business permit from the new Waters-Pierce Oil Company. Still, Smith was annoyed because Pierce had not even bothered to change his company's name. Hardy required Pierce to sign an affidavit swearing that Waters-Pierce was not a member of the oil trust. He then granted the new permit.[46] Attorney General Smith had no way of knowing of the secret plan to maintain Standard Oil stock control of Waters-Pierce, but he possessed adequate information to withhold the permit if he had been genuinely interested in breaking the oil monopoly in Texas. At the Chicago Antitrust Conference, Smith had declared: "I take the position that no corporation can do business in another state without the consent of that state."[47] But Smith's treatment of Waters-Pierce stood in stark contrast to his public pronouncement. In a May 15 letter to Johnson—dated two months after the Supreme Court ruling sustaining the ouster of Waters-Pierce and only sixteen days before he approved the new permit—Smith wrote:

. . . I have today received a letter from Luling, Texas, in which it is stated that the company there is doing business like it used to do; that they sell to all the people alike through their agent and afterwards allow some a rebate of from 1 to 2 cents a gallon, thus giving one merchant an advantage over others. I think this should be looked after by Mr. Pierce.[48]

The Waters-Pierce Oil Company had been expelled from Texas because of illegal rebating. The attorney general had proof that the company still granted rebates. He professed to believe that "no corporation can do business in another state without the consent of that state." Yet he advised Hardy to issue a business permit to a rechartered Waters-Pierce.

For Joe Bailey, 1900 was a busy year. In addition to buying the Grapevine Ranch and arranging the readmission of Waters-Pierce, he won a seat in the United States Senate. Bailey campaigned vigorously early in the year and scored an impressive series of county primary victories. His opponent, incumbent Senator Horace Chilton, became ill and was unable to meet his campaign schedule. Confronted with Bailey's primary victories, Chilton suddenly withdrew from the race on April 17. Since Bailey faced no other opponent, he had in effect won the Democratic nomination, and in Texas that meant certain victory.[49]

Bailey's successful campaign for the Senate guaranteed that his association with Henry Clay Pierce would receive public attention. Controversy over the readmission of Waters-Pierce erupted at the Texas State Bar Association meeting at Galveston in July 1900. Judge James L. Autry delivered an address to the association on out-of-state corporations, in which he admitted that the readmission of Waters-Pierce allowed that company to continue operations in Texas "without jar or friction." Nevertheless, Autry asserted that the business permit could not have been legally denied, although the secretary of state certainly could have delayed action.[50]

Jim Hogg heard Autry's address and took strong exception to his conclusions. He told the meeting that the secretary of state should have withheld the permit and forced the company to go to court to gain readmission. He portrayed the Waters-Pierce reorganization as a transparent fraud, observing that the company "didn't even change the mules which were drawing its wagons" to gain readmission to the state.[51] Attorney General Smith and other notables of the Texas bar ridiculed Hogg's lack of legal sophistication. They claimed that the old Waters-Pierce Oil Company had been legally dissolved, and that a new concern, owned exclusively by Henry Clay Pierce, had obtained a valid charter.[52] But Hogg remained unconvinced.

The scene of battle then shifted to the Democratic state convention that convened at Waco on August 8. Bailey was scheduled to receive the nomination for the Senate at this meeting, and Governor Sayers and Attorney General Smith both sought renomination. These politicians understood the potential damage of the Waters-Pierce affair. Therefore, on the first day of the convention Bailey took the floor to defend his actions. He proclaimed that he had dealt with Pierce merely as a personal favor for his old friend Dave Francis. Pierce himself had sworn to Bailey that the oil trust had "never owned a controlling interest" in Waters-Pierce. Bailey noted that he had declined to be Pierce's lawyer but added, "I happened to be at the capitol at that time [May 1] and talked with the Attorney General."[53] Neglecting to mention that Pierce had loaned him $3,300, Bailey concluded his oration to the enthusiastic cheers of the assembled delegates and spectators.

Hogg then addressed the convention. He proclaimed that "the material question presented by the repermission of the Waters-Pierce Oil Company to do business in this state is: Shall Texas, or the trusts control? [Wild Cheers]."[54] He questioned the propriety of allowing the secretary of state to nullify the laws of Texas and the decree of the Supreme Court, and he condemned both Smith and Hardy for their hasty and careless action. Despite Hogg's rhetoric, the Bailey forces maintained control of the convention. A move to censure Bailey was defeated, and he was nominated to the United States Senate. Sayers and Smith also won renomination.

Why did Texas Democrats refuse to support Hogg on the Waters-Pierce issue? His proposition that the new charter had not changed the company's everyday operations was irrefutable. Yet Hogg lacked several key facts that would have proved his case. He did not know of the covert Standard Oil plan to retain 2,748 shares of Waters-Pierce stock. Attorney General Smith kept secret the continued rebating by Waters-Pierce, and Bailey successfully hid his $3,300 no-collateral, no-interest loan from Pierce. Moreover, Texas Democrats placed Hogg's position in the Waters-Pierce controversy in its political context. Hogg had strongly opposed Bailey's nomination for senator, actively campaigning against him in the primaries. Hogg had also opposed Governor Sayers's election in 1898.[55] Texas Democrats thus knew that Hogg wanted to use the Waters-Pierce affair to embarrass his political enemies. They understandably viewed his convention performance as a rearguard action by the leader of a declining political faction.

Texas Democrats also knew that Hogg was an inconsistent opponent of monopoly. While battling trusts at home, Governor Hogg had toured the Northeast in the summer of 1894 reassuring corporate investors that Texas was a safe place for their money. After leaving office in January 1895, he had immediately established a lucrative practice as a railroad attorney. Only forty-nine years old in 1900, Hogg consistently rejected consideration of a seat in the United States Senate for financial reasons. He wrote to one political supporter: "I unhesitatingly refuse to become a candidate or accept the office [of Senator] for I am financially unable to do so. My practice is good, and I am fast retrieving the losses caused by my eight years service [as attorney general and governor]."[56] In

1898 Hogg again toured the Northeast promoting corporate investment in Texas. Given this record, the convention delegates viewed
Hogg's impassioned pleas for a war on the oil monopoly with considerable skepticism.[57]

The state legislature, empowered to elect Bailey to the Senate,
convened in January 1901. Although Bailey's supporters dominated the body, his opponents were determined to put up a fight.
Representative David McFall, introducing a resolution calling for
an investigation of the Waters-Pierce affair, branded the company's relicensing a fraud and claimed that the act had "the passive
assistance of certain State officials and the active assistance of Congressman Joseph W. Bailey."[58] Bailey urged his supporters to back
the investigation. He knew that his opponents did not possess the
most damaging evidence against him; he wanted to crush his
detractors and prematurely end the controversy.[59]

The house named a committee of seven to examine the evidence,
and hearings began on January 14, 1901. Claiming illness, McFall
failed to testify before the committee in support of his charges.
Two witnesses offered evidence of the anticompetitive practices of
Waters-Pierce, but they were unable to prove that the company was
a member of the trust.[60] While Bailey's opponents floundered,
several witnesses defended him, including "Honest Tom" Smith
who repeated that the state could not have legally denied Waters-
Pierce readmission. Bailey also took the stand on his own behalf.
Again omitting any reference to his $3,300 loan, he repeated that he
had not received a fee for his services for Henry Clay Pierce. He
then added: "I would not have hesitated one second about accepting employment from him and taking a fee; . . . I am a lawyer and I
sincerely hope I will never be such a political coward as to fear to
defend any man or any business when I think their rights are assailed
unjustly, or with unnecessary severity."[61]

Bailey's performance convinced the committee members. They
unanimously exonerated the senator-to-be, and their report round-
ly denounced Bailey's critics. One member of the committee refused to sign the report because he felt that the savage attack on
Bailey's critics would inhibit frank criticism of public officials in
the future. But the majority of the Texas House felt no such inhibi-
tions. On January 23, they approved the committee report by a

vote of 87 to 25. On the same day, the full Texas legislature gave Bailey 137 of 141 votes for United States senator.[62]

The Waters-Pierce Oil Company continued its lucrative operations in Texas. Capitalized at $400,000, it reported the following total profits for the five-year period after readmission:[63]

	Profits	Profits as Percentage of Capitalization
1900	$1,813,032.88	453%
1901	$1,987,184.33	497%
1902	$2,001,203.18	500%
1903	$2,699,818.68	679%
1904	$2,790,981.87	672%

Senator Bailey continued to work for Waters-Pierce. In February 1901 a motion was introduced in the Texas Senate to instruct the attorney general to sue the company for damages under the antitrust law. That same month, McFall introduced a bill in the Texas House to revoke the company's business charter. On March 1, 1901, Bailey received another $8,000 loan from Pierce. On March 11 he rushed back to Austin from Washington and successfully supervised the defeat of these two measures.[64]

Meanwhile, on the Gulf Coast near Beaumont, Texas, events of far greater significance were unfolding. On January 10, 1901, Captain Anthony Lucas drilled eight hundred feet into a barren mound of earth called Spindletop and struck oil. Uncapped for over a week, the Lucas gusher spewed forty thousand barrels of crude into the air each day. Houston, seventy-five miles away, became an oil town overnight, and wildcat drilling became the rage in Texas. Almost five hundred drilling companies were chartered in the state in 1901 alone. The Gulf Oil Corporation and Texaco, both destined to become multinational titans, got their start in this field. In its first year the Gulf Coast area produced 3.7 million barrels of oil. By 1905 annual output had climbed to 36.5 million barrels, a full 27 percent of total United States production.[65]

From New York, these events somehow seemed less significant. The Standard board of trustees judged Spindletop "a flash in the pan." Already producing oil in widely scattered locations, they decided not to enter the new field in force.[66] Even if the Standard leaders had committed themselves fully to Texas, they probably could not have controlled the region. The Gulf Coast strike was too vast and too sudden for any single organization to handle. Nature had succeeded where the law had failed. Spindletop brought competition, if not chaos, to the Texas oil industry.

While savoring the excitement of an oil boom, Texans remained wary of the machinations of the Rockefeller organization. Waters-Pierce, despite a swarm of new competitors, defiantly conducted a flourishing and highly profitable operation in the state. Consequently politicians kept an eye on the firm. In 1901 Attorney General Charles K. Bell filed suit against Waters-Pierce as soon as he discovered that its officials had neglected to sign an affidavit—required annually from all businesses in Texas—professing compliance with the antitrust law. But Bell had to drop the suit when company officials completed the necessary papers.[67] That same year, a county attorney attempted to extract penalties from Waters-Pierce under the 1899 Texas antitrust act. This effort developed into a minor disaster. The trial court acquitted the company, and the Texas Court of Civil Appeals declared the antitrust law unconstitutional because it exempted agricultural producers. In 1903 the state legislature had to pass a new antitrust statute, minus the offending provision.[68] Rearmed and undaunted, Texas trustbusters awaited their next opportunity to strike at Standard Oil.

The Waters-Pierce issue remained politically explosive in Texas primarily because it involved Senator Joe Bailey, who, by early 1906, was a rapidly rising star in the Democratic party. At the youthful age of forty-two, Bailey had emerged as a leading figure in the United States Senate. He recently had received widespread praise for his role in the passage of the Hepburn Act. *Current Literature,* for example, portrayed him as the "foremost Democratic leader in America," and Texas Democrats talked extravagantly of their erratic senator as the party's next presidential nominee.[69]

Bailey was at least as controversial as he was influential. In the July 1906 issue of *Cosmopolitan Magazine,* David Graham Phillips attacked the Texas lawmaker in the fifth installment of his series, "The Treason of the Senate." Phillips developed the sensational, and unsupportable, thesis that Senator Nelson W. Aldrich of Rhode Island presided over a merger of key senators from both parties, including Bailey, who worked to support "the interests" and despoil the people. Drawing on evidence recently uncovered by the attorney general of Missouri, Phillips exposed Bailey's role in the readmission of Waters-Pierce. He also noted the Texas senator's flamboyant life-style, supported by substantial law fees from large corporations. Although he occasionally distorted the facts, Phillips forcefully portrayed Joe Bailey's formidable talent for converting political power into personal wealth.[70]

Because of the Phillips article, Bailey worked diligently during the 1906 political campaign to solidify his position in Texas. He was unopposed for reelection to the Senate in the July state primary, and he went to the Democratic State Convention in August wielding substantial power to influence the outcome of the close gubernatorial race, in which three reform-minded candidates —all advocating strict antitrust enforcement—had a chance to win. Bailey was careful; he kept his preferences secret early in the convention and then threw his support behind Thomas M. Campbell at just the right moment. Campbell won the nomination the following day, and Bailey gained the support and gratitude of the next governor of Texas. He would need it.[71]

By mid-1906 extensive publicity, highlighted by the Phillips piece, had made the oil trust's continuing control of Waters-Pierce common knowledge. True to Texas tradition, Attorney General Robert Vance Davidson was pursuing the preliminaries of yet another antitrust action against the St. Louis-based firm. Davidson, however, had more than law enforcement on his mind. He was politically ambitious, and he knew that the expulsion of Waters-Pierce would have tremendous political impact. Better than that, conclusive proof of Bailey's long-rumored connection with Standard Oil would be a political master-stroke. Davidson's management of the antitrust effort against the Standard combination would reveal that his goals were essentially political, not legal.

The attorney general's unfortunate involvement with J. P. Gruet indicates the extent of his political motivation. Gruet was an ex-Waters-Pierce auditor, who had worked in the St. Louis office and had thus gained access to confidential company files. In 1904 the Standard board shook up the Waters-Pierce management, and one of the new men from New York observed that Gruet was an alcoholic. Shortly after this discovery Henry Clay Pierce received word that his auditor was expendable. In April 1905 Gruet found himself in an inferior position in another firm—the Pierce Investment Company. A year later, Gruet suddenly resigned, claiming that Henry Clay Pierce owed him $28,000 in back wages.[72]

Since 1903 Gruet had been secretly taking confidential documents from the company files concerning Senator Bailey and the readmission controversy. In April 1906 he reproduced and altered these documents. Two months later Gruet's agent wrote Senator Bailey, suggesting that the senator convince Pierce to turn over $28,000 to his former employee. The agent added that Gruet possessed "documents that looked very bad" for the Texas lawmaker. Furthermore Gruet would not hesitate to use this evidence to recover his money. Bailey did not respond to this threat, so Gruet turned to Attorney General Davidson.[73] On August 26, 1906, the alcoholic auditor made an agreement with the state of Texas that granted him one-third of the 25 percent prosecuting attorney's commission provided by the antitrust law. Since the state planned to request penalties of $5 million in the case, Gruet could have received the incredible sum of $400,000 for his altered documents.[74]

Davidson was obviously unaware of Gruet's forgeries. The record is unclear as to whether he knew of the blackmail attempt against Bailey.[75] Nevertheless the attorney general knew about Gruet's background. He should not have offered vast amounts of state money for evidence supplied by a witness of that caliber. In fact, Gruet's evidence was not even needed to expel Waters-Pierce from the state; that company's connection to the oil trust was already public knowledge. But Davidson wanted Gruet's documents to tie Bailey to the reentry fraud. He desired to place Bailey in the middle of the antitrust suit in order to increase the political impact of the case.

On September 10, 1906, news from St. Louis jolted Texas. Henry Clay Pierce took the stand in a Missouri antitrust hearing and publicly admitted that Standard Oil still controlled his company. Although Pierce revealed nothing new, the spectacle of the oil baron personally explaining how he had evaded Texas law for years had great impact in the state. Five days later Pierce testified in a suit brought by Gruet to recover his back wages. This time the oil tycoon disclosed that in 1905 he and several other prominent St. Louis businessmen had hired Joe Bailey to handle their $6.5 million investment in the Tennessee Central Railroad. Pierce's revelation of his recent dealings with the enterprising senator intensified the political uproar and led to spontaneous public meetings throughout Texas to protest Bailey's reelection.[76]

On September 20, at the crest of the Texas reaction to Pierce's testimony, Attorney General Davidson filed suit against Waters-Pierce, demanding cancellation of the company's business permit and a $5 million fine. Davidson also charged that "Henry Clay Pierce came into the State of Texas with a large sum of money" and thereby "secured a permit to do business."[77] At the pretrial hearing the judge struck out Davidson's allusions to Pierce's financial entanglements, observing that such charges injected an "improper element" into the case. Davidson had to amend his petition, but he defiantly served notice to the Waters-Pierce attorneys that they would have to produce all records relating to Senator Bailey and the readmission of their company.[78]

Apparently seeking maximum publicity, Davidson then issued an open letter to Bailey based on Gruet's documents. He charged the senator with accepting the following fees for services rendered to Henry Clay Pierce: $3,300 on April 25, 1900; $1,500 on June 15, 1900; $200 on November 23, 1900; $8,000 on March 1, 1901; and $1,750 on March 28, 1901. According to the attorney general, these payments appeared on the Waters-Pierce books marked "account of Texas cases."[79]

Senator Bailey attempted public explanation. The embattled public servant recollected that the $8,000, $3,300, $1,750, and $1,500 amounts were installments of some personal loans from Pierce, loans that he had repaid. He knew nothing of the $200 payment. Bailey observed that he had never seen the Waters-Pierce books so he had no idea how Pierce had recorded the transactions.

The senator then demanded that Davidson reveal his source of information. Because of the confused nature of some of the attorney general's accusations, Bailey charged that Davidson had been duped by a forger.[80]

This heated exchange elicited widespread comment. The *Dallas Morning News,* an influential paper that had long supported Bailey, "did not consider Senator Bailey's explanation satisfactory." The editorial went on to conclude that Bailey's "influence and usefulness have been so sadly impaired by the charges against him and by his own responses and explanations that the Democracy of Texas should unite upon some other man for the high place he now holds."[81] Many other papers abandoned Bailey. Substantial numbers of Texas Democrats, led by William A. Cocke, a state representative, opposed the senator's reelection, and ambitious politicians contemplated last-minute entry into the Senate race. But Bailey's faltering campaign received a boost when Governor Campbell repaid his convention debt by refusing to join the fight against the senator.[82]

When the Texas legislature convened in early January 1907, there were calls in both houses for an investigation of Bailey's role in the Waters-Pierce affair. Cocke, leader of the anti-Bailey forces in the Texas House, submitted a bill containing forty-two formal charges against the senator, and the speaker of the house appointed a seven-member committee to look into these charges. The Texas Senate also formed a committee of seven to join in the investigation.[83] But Bailey's opponents were unable to block his scheduled reelection on January 22. Although thirty-one candidates received nominations, Bailey easily won the election, collecting 108 of 147 votes. Following this triumph, Bailey confidently notified the legislature that he would resign if his accusers proved any of the forty-two charges against him.[84]

The investigators failed to connect the senator with a distinctly illegal act. Gruet's performance on the witness stand was a disaster. The former auditor's alcoholism, his salary dispute with Pierce, his blackmail attempt against Bailey, and his contract with Davidson to share the attorney's fee were all exposed, and they shattered his credibility.[85] Furthermore, other witnesses proved that Gruet had altered key documents. Nevertheless Bailey's detractors were able to expose the senator's extensive, unethical forays. Bailey reluctantly admitted that he had received two no-interest loans totaling $12,800 from Henry Clay Pierce while he helped Waters-Pierce get back into Texas. The senator also disclosed lucrative arrangements with other corporations. During his first term in the United States Senate (1901-1907), he had received fees of $200,000 from the Kirby Lumber Company, $100,000 from the Tennessee Central Railroad, and $5,000 from Security Oil. This last fee was significant—despite its modest size—because Security Oil was part of the Standard combination.[86]

At the conclusion of the investigation, the majority of the Texas Senate, which was strongly pro-Bailey, feared that their committee would issue a report critical of the senator. They therefore dissolved the potentially troublesome body and voted for a complete exoneration of the senator. The house, with more anti-Bailey members, allowed its committee to file reports. Four of the seven members declared Bailey innocent of each of the forty-two counts, but three dissendents submitted a minority report, declaring: "at the most the evidence shows a course of dealing on his part deemed by many to be inconsistent with sound public policy."[87] For the second time in six years, the Texas legislature had rallied behind Joe Bailey, despite overwhelming evidence that he consistently placed financial gain above political duty.

Davidson had failed to drive Bailey from the senate, but he still had an excellent opportunity to drive Waters-Pierce from the state. His antitrust case went to trial before a jury in the district court at Austin in May 1907. After the catastrophe before the investigating committee, Davidson wisely decided not to place Gruet on the stand. Information gathered in the antitrust probe supplemented by Henry Clay Pierce's recent St. Louis testimony proved more than adequate to obtain a conviction. On June 3, 1907, the jury found the company guilty on eighteen counts, all relating to the Waters-Pierce tie to Jersey Standard. These convictions carried fines totaling $1,623,000. The company also lost its business permit.[88]

The Texas Court of Civil Appeals upheld the conviction, and the Texas Supreme Court refused to grant Waters-Pierce a writ of error without opinion. On January 18, 1909, the United States Supreme

Court handed down its decision in the case. Justice William R. Day, for a unanimous court, sustained the fine and ouster. Day rejected defense arguments that the fine was so excessive that it violated due process of law. He noted that Waters-Pierce had conducted a lucrative and illegal business in Texas for many years, now owned property valued at $40 million, and frequently paid annual dividends as high as 700 percent. The fine was not excessive considering the scope of this enterprise.[89] Attorney General Davidson, who appeared before the Supreme Court on behalf of the state, had won a substantial victory. He would run for governor in 1910 on the strength of this prosecution.[90]

Although the suit was a notable political and legal success, its economic impact was insignificant. After Spindletop, monopoly was no longer an issue. Hence Davidson's primary goals were to exact a heavy fine for past violations and, more important, drive the oil trust from the state. He clearly achieved his first objective; the courts awarded Texas $1,640,000 in penalties. The fine was a great political achievement for Davidson, as well as a personal bonanza, a provision of the Texas antitrust law granted the prosecuting attorneys a 25 percent share. But the effectiveness of the fine as punishment for antitrust violations is less certain. Waters-Pierce and Jersey Standard shared the penalty. Waters-Pierce, which paid 32 percent of the fine, had averaged profits of over $2 million in each of the first five years of the century. Jersey Standard, which paid the remaining 68 percent, averaged annual earnings of $68 million in the 1900-1906 period.[91] Moreover, the economic impact of fines on giant corporations is questionable. As the New York *Press* remarked: "Every one of these fines is not a punishment for the monopoly, but for the consumers of the monopoly's product. The bigger the fine the more the public has to pay for oil."[92]

Thus Davidson's only truly meaningful goal was to drive Standard Oil from Texas, yet the combination remained in the state long after the Waters-Pierce ouster. By 1906 more than one Standard company operated in Texas. In 1898 the National Transit Company, a Standard affiliate, provided capital for the organization of the Navarro Refining Company. This firm produced and stored oil, and it completed a large oil refinery in 1899. That same year, the National Transit Company formed the Corsicana Petroleum Company, which, despite its name, concentrated on the development of natural gas fields. Still another Standard-backed firm, the Security Oil Company, constructed a large oil refinery near Beaumont, Texas, in 1903.[93] These Standard firms were among the largest oil companies in Texas. In 1906 Gulf Oil and Texaco, the leading independents, refined 3.9 and 0.9 million barrels of oil, respectively. In that same period Security and Navarro refined approximately 1.6 and 1.1 million barrels per year.[94] Despite Davidson's strong anti-Standard posture, these Standard companies continued to operate in Texas until Senator Bailey injected them into the political struggle.

In October 1907 Bailey publicly noted the attorney general's failure to expel all the Standard companies from Texas. The senator pointed out that the connection between these companies and the oil trust had been clearly established during the Waters-Pierce investigation. Bailey challenged Davidson to drive out all the Standard companies.[95] Stung by Bailey's attack, the attorney general filed suit against Navarro Refining, Security Oil, and Union Tank Line Company—which owned and operated oil tank cars—on November 6, 1907. The state charged that these three companies were all owned, directly or indirectly, by Jersey Standard and that they worked in harmony to lessen competition substantially in the petroleum industry. Davidson asked for forfeiture of the Navarro and Security charters, as well as an absurdly high fine of $55 million.[96] Despite this belated display of zeal, Davidson's failure to act without prompting indicates that he was more interested in attacking a politically controversial target like Waters-Pierce than pursuing the routine litigation necessary to carry out his professed goals.

After a delay of almost two years, the case finally went to trial. Standard lawyers admitted the guilt of the three firms. On October 27, 1909, the district court judge at Austin revoked the Security and Navarro charters. Union Tank Line, which had been operating in Texas without a business permit, had previously stopped doing business in the state. The judge placed the Texas property of all three companies in receivership and fined them a total of $154,000.[97]

The culmination of Davidson's crusade against the oil trust occurred at the Driskill Hotel in Austin on December 7, 1909, when the court-appointed receiver sold the property of the three companies to John Sealy, who headed a partnership composed of leading members of the Standard combination. Assistant Attorney General Jewel P. Lightfoot, who represented Davidson at the affair, approved the sale, despite frantic pleas from those present that he require Sealy to disclose his partners.[98] John Sealy & Company ran the Security and Navarro refineries and pipelines for over a year with no changes in management or operations. Then, in April 1911, Sealy organized the Magnolia Petroleum Company to take control of these businesses. John D. Archbold and Henry C. Folger, Jr., both prominent members of the Standard board of trustees, owned 85 percent of the stock in this new enterprise.[99]

The receiver for the Waters-Pierce Oil Company also sold that firm's property at the Driskill Hotel on the same day. Sam Fordyce, a St. Louis businessman and old friend of Henry Clay Pierce, was the buyer. Again, Lightfoot appeared to have no interest in who acquired control of the company. Shortly after the sale, Pierce and Fordyce formed the Pierce-Fordyce Oil Association. Pierce owned a controlling interest in this nominally new company that carried on the same business as Waters-Pierce. Because Pierce still operated the Waters-Pierce Oil Company in other southwestern states and because Jersey Standard still owned 68 percent of that company, the Pierce-Fordyce Oil Association remained indirectly tied to the Oil Trust.[100] Hence the Texas attorney general's office approved the return of the property of Security Oil, Navarro Refining, and Waters-Pierce to the control of the Standard organization.

Although Attorney General Davidson caused the reorganization of several companies, he did not expel the Standard combination from Texas. He failed because he consistently pursued political advantage at the expense of the law. He unnecessarily injected the Bailey issue into the Waters-Pierce suit. He sought spectacular fines, despite their questionable economic impact. He cared more about destroying the Waters-Pierce name than policing the actual ownership of the company. The attorney general's preoccupation with politics is the only explanation for his needless purchase of Gruet's doctored evidence, his failure to move independently against Security Oil and Navarro Refining, and his negligent supervision of the Driskill Hotel receiver's sale. Successful antitrust litigation against the powerful oil trust was extremely difficult

under the best of circumstances. Davidson's unbounded political
ambition made it impossible.

On the surface, the prospects for successful antitrust action
against Standard Oil were radically different in Ohio and Texas.
Ohio was a conservative, industrial state dominated by the Republican party, while Texas remained primarily agricultural and inordin-
ately proud of its antimonopoly tradition. Yet trust busting proved
no more effective in the Southwest than it had in the North. The
criminal conspiracy indictments of 1894 simply embarrassed the
Hogg and Culberson administrations. Texas actually ejected
Waters-Pierce through the 1895 quo warranto suit, only to have
this victory bargained away by Joe Bailey and the Sayers administration. Attorney General Davidson casually discarded the fruits of
his antitrust forays at the Driskill Hotel receivers' sale. In all, state
officials filed seven separate actions against agents and affiliates of
the Standard organization. Counting appeals, fifteen decisions, including three by the United States Supreme Court, resulted. But
none made much difference to the state's petroleum industry.[101]

Economic reality had imposed a fundamental dilemma upon the
state's politicians. They had grown up with a strong antimonopoly
tradition, and political pressure from the Populist party led them to
embrace that tradition. They therefore passed a broad antitrust law
and initiated several antitrust suits. To a certain extent, they were
sincerely committed to the cause. But the Texas politicos knew that
long-term development for the state meant industrialization. The
recurring depressions in agriculture made them particularly sensi-
tive to economic conditions. To industrialize Texas had to attract
large investments from eastern corporations, and in this period
these corporations were rapidly becoming massive nationwide organizations. Bailey unwittingly outlined this dilemma in his testimony before the Texas legislature in 1901:

I am as much opposed to trusts as any man, and am as ready to enact drastic measures to suppress them; but I have sense enough to understand that
the surest way to render our opposition to the trust ineffectual is for us to
blindly attack legitimate and useful enterprises simply because they happen
to be successful.[102]

Before the rise of the great trusts and the passage of the antitrust law, the Texas tradition of antimonopoly oratory flourished untested. The Standard Oil cases offered Texas politicians a concrete opportunity to drive the oil trust from the state, but their words proved hollow.

The central significance of the Waters-Pierce episode in Texas is not that Joe Bailey helped the devious Henry Clay Pierce get his company readmitted or that Attorney General Davidson allowed Standard interests to buy back into the state; corrupt politicians and unscrupulous businessmen are an old story in American history. Rather, the primary fact is that the Texas political establishment twice examined the Waters-Pierce issue and both times approved Bailey's election to the Senate. Texas Democrats, frightened by depression and enticed by the prospect of prosperity through industrial development, were incapable of implementing their extravagant antimonopoly pronouncements. In Ohio, Standard Oil exerted an open and overpowering influence on state politics. In Texas, the influence was more subtle but no less effective.

The public discussion of these events created the wholly erroneous impression of repeated triumph by the state over Standard Oil. As in Ohio, the papers focused on clear-cut, dramatic events like the million dollar fine and the ouster orders. The minutia of corporate reorganization and the drawn-out process of appeal, although frequently more important, commanded less attention. Political attacks on Senator Bailey cast some light on the underside of these cases. But Bailey, who repeatedly rallied the Austin political establishment to his defense, outwitted his detractors and blunted the impact of their criticism. Far more important than the press, Texas politicians obscured the essential facts of this litigation. The leading figures on both sides of these controversies—Bailey, Sayers, and Smith, as well as Hogg, Culberson, and Davidson—were all ostensibly devoted to the Texas antitrust tradition, and they all labored mightily to get their message out to the voters. Collectively, they succeeded to a remarkable extent in disguising the ineffectuality of their antimonopoly activities. Texas acquired a reputation as a foe of Standard Oil that persists to this day. In fact, some leading recent surveys of the oil industry have portrayed the state as an effective adversary of the Rockefeller combination.[103]

(3)
The Antitrust Contagion

During the 1890s antitrust activity against Standard Oil was essentially confined to Ohio and Texas. But after the turn of the century trust busting at the state level assumed the proportions of a national crusade. Prosecutors representing eight additional states and the Oklahoma Territory competed with their colleagues from Texas and Ohio for the attentions of the Standard defense attorneys.

Reform-minded writers and orators, who for years had focused public attention on the Rockefeller organization, made Standard the very embodiment of monopoly and thus laid the groundwork for this rash of lawsuits. In 1894 Henry Demarest Lloyd won critical acclaim for *Wealth Against Commonwealth,* a work that portrayed Standard Oil as a serious threat to the foundations of American society. The broadest and sharpest attack, however, came after 1900. Ida Tarbell led the new wave of oil trust critics. Her nineteen-article series, "The History of the Standard Oil Company," which appeared in *McClure's* from 1902 to 1904, enflamed the public's long-standing hostility toward the combination as nothing before had. In Washington, Theodore Roosevelt, who became president in 1901, played upon and compounded this antagonism by repeatedly picturing Standard as the nation's outstanding example of an evil trust. John D. Rockefeller became a prime target for monopoly haters. After 1902 Rockefeller routinely received threats on his life. During this period the oil baron kept a revolver beside his bed at night, and his pastor hired Pinkerton detectives to mingle with the crowd that gathered each Sunday to watch the devout multimillionaire attend church.[1]

Concurrent with hostile publicity, a series of investigations established an exceptionally solid factual basis for legal action against the combination. As early as 1879 the Hepburn Committee of the New York State Assembly had examined Standard's practice of extracting rebates and special rates from the railroads. And in 1888 both the New York Senate and the United States House of Representatives had looked into Standard's position of unchallenged supremacy in the petroleum industry. Like the negative publicity, the investigations became more intense after the turn of the century. In 1900 the United States Industrial Commission released a massive thirteen-volume report filled with damaging material on Standard Oil. The new federal Bureau of Corporations filed an extensive report on the interstate transport of petroleum products in 1906. The following year the bureau issued a two-volume survey of all aspects of the oil business, while the Interstate Commerce Commission published a report on the transportation phase of the industry. All of these emphasized Standard Oil's domination of the industry and its frequent resort to anticompetitive practices.[2]

But writers and investigators were not the prime movers in the antitrust crusade. The explosive growth of the oil business itself was the underlying source of the proliferating litigation. As oil became a central fact in the daily lives of the American people, the possibilities for economic conflict and the potential political rewards for those who could resolve those conflicts in the public interest increased dramatically. In the first decade of the twentieth century, annual oil production in the United States more than tripled from 63 million to 209 million barrels. Most of the new oil came from virgin fields in widely scattered sections of the country. The old Appalachian and Ohio fields gradually declined in productivity, while Illinois, Kansas, Oklahoma, the Gulf Coast, and California provided vast new resources. These new areas provided opportunity for independent producers and refiners, who frequently clashed with Standard Oil and were more than willing to take their disputes to court. In Kansas, for example, agitation by independent oil producers led directly to a state antitrust suit against the oil trust.[3]

The uses of petroleum were changing just as rapidly as the sources. In the nineteenth century, illuminating oil was by far the most important product of American refineries. This commodity alone accounted for over 80 percent of refinery output in the 1880s. But after 1900 fuel oil, gasoline, and lubricants became increasingly important, comprising about 60 percent of refinery runs in 1910. Fuel oil was used to run ships, trains, and factories. More important for the future, gasoline powered the automobiles that were rapidly spreading throughout the country. And in 1903 Standard agents were even at Kitty Hawk, North Carolina, offering fuel to the Wright brothers. These new uses of petroleum products created vast numbers of customers who were vitally concerned with Standard prices and marketing practices. Like the independent producers and refiners, these consumers were willing to support antitrust activity to keep the oil trust in line. In a very real sense, the expansion of the petroleum industry precipitated the antitrust crusade against Standard Oil. The court battles in Tennessee, Kansas, and Missouri reveal the essential character of the resulting wave of litigation.[4]

Early in the new century Tennessee officials began a modest antitrust effort that provides a brief, clear example of how this most delicate legal process actually worked. Tennessee, like Texas, had a long-standing antimonopoly tradition. The state constitution of 1870 declared that "perpetuities and monopolies are contrary to the genius of a free state, and shall not be allowed."[5] The Tennessee legislature was among the first in the nation to pass an antitrust law, approving their bill on April 6, 1889. Furthermore, the legislators demonstrated their continuing concern with the trust problem by strengthening the law in 1891, 1897, and 1903.[6]

The 1903 version of the Tennessee statute proclaimed that "all arrangements, contracts, agreements, trusts, or combinations between persons or corporations made with a view to lessen, or which tend to lessen full and free competition" were against public policy and thus void.[7] The law penalized corporate violators through forfeiture of charter or business permit, and individual offenders were subject of fines from $100 to $5,000 and prison terms from one to ten years. The state attorney general or district attorneys could begin suits on behalf of the state. Individuals also could bring suit to recover damages inflicted by a combination in restraint of trade.[8]

Less than a year after this measure took effect, Standard Oil of Kentucky, which marketed petroleum products for the combination in Tennessee and several other Mississippi Valley states, was before the courts as an antitrust offender. Kentucky Standard was originally known as Chess, Carly & Company, the same firm that had battled Henry Clay Pierce for customers in the 1870s. In 1886 the Standard board of trustees bought out all minority stockholders and renamed the company. By the turn of the century, Kentucky Standard had captured practically the entire Tennessee market. The company shipped its products into the state by rail from Standard refineries in Cleveland and Whiting, Indiana. Kentucky Standard established "agencies" at most major towns in the state where products were unloaded from railroad tankcars and stored in bulk. From there, the oil was placed in tank wagons for distribution to retail merchants or directly to consumers.[9]

Kentucky Standard had such an agency at Gallatin, Tennessee, a town located in Sumner County about twenty miles northeast of Nashville. C.E. Holt was the regional sales representative for Gallatin and the surrounding area. J.E. Comer, his immediate supervisor, was Kentucky Standard's superintendent for north central Tennessee. The company sold coal oil in the region for the reasonable price of $.135 per gallon, but product quality was poor, and complaints were numerous.[10]

In October 1903 a salesman for the Evansville Oil Company visited Gallatin. Evansville Oil was an independent firm with headquarters in Indiana and a refinery at Oil City, Pennsylvania. The salesman offered Gallatin consumers high-quality Pennsylvania coal oil for $.145 per gallon, only a cent per gallon more than the inferior Standard product. Since many Tennesseans were dissatisfied with the quality of Standard coal oil, the Evansville representative had little difficulty securing orders for a railroad tankcar of his product.[11]

Standard agents quickly learned of the Evansville Oil contracts through their informants, who worked for the railroads. The agents relayed this information to Comer, and he promptly ordered Holt "to go to Gallatin and hold his trade, and procure the orders that had been given to the Evansville Oil Company countermanded.''[12] Holt visited the Evansville customers and offered free oil to those who would cancel their orders. Several merchants agreed, and they received a total of three hundred gallons of free oil for their cooperation. One of the merchants, S.W. Love, sent a telegram to Evansville Oil, at Holt's expense, canceling the orders. But cancellation came too late. The oil was already in transit.[13]

Comer then informed the comptroller of Tennessee that Evansville Oil was shipping petroleum into the state. Comer took this action to guarantee that Tennessee would tax the shipment. As originally contracted, the Evansville Oil shipment was not taxable because it was a direct interstate transaction. But since a substantial part of the order had been canceled, much of the oil would have to be stored at Gallatin and subsequently resold, thus becoming liable to state taxation. To ensure that Tennessee would collect those taxes, Holt also told the county clerk at Gallatin about the oil shipment.[14]

When Evansville Oil's cargo arrived at Gallatin, Love and the other merchants refused delivery. The company therefore had to store the oil and find new buyers, which meant storage expenses and state taxes. Evansville Oil consequently lost a substantial sum on the transaction and decided to abandon further attempts to market oil in the area. After the independent oil company had withdrawn, Kentucky Standard increased the price of its low-quality coal oil to $.145 per gallon.[15]

On December 22 and 23, 1903, newspapers in Nashville and Gallatin featured accounts of Kentucky Standard's successful campaign against Evansville Oil. The local district attorney responded to the resulting public outcry by quickly investigating the matter and presenting his evidence to a grand jury. On February 3, 1904, that body returned indictments against Holt and Kentucky Standard, charging that the company and its agent had conspired with Love "to lessen and destroy full and free competition" by procuring the cancellation of the Evansville Oil contracts through the gift of free oil. Holt and Kentucky Standard were tried at the circuit court of Sumner County, and the jury found both defendants guilty. The court fined the company $5,000 and Holt $3,000.[16]

Kentucky Standard then appealed the case to the Tennessee Supreme Court. On February 26, 1907, Justice John K. Shields delivered the opinion for a unanimous court. He rejected Kentucky

Standard's argument that the Tennessee antitrust act unconstitutionally infringed on the congressional commerce power. He upheld Holt's fine. But he overturned the company's financial penalty because the wording of the antitrust law did not allow fines against corporations. The statute, he declared, prescribed only forfeiture of charter or revocation of business permit as penalties for corporations.[17]

Shield's ruling was a reasonable interpretation of the statute. It was also a major opportunity for the state. A $5,000 fine against Kentucky Standard was a meaningless gesture. But Shield's decision allowed Attorney General Charles R. Cates to file a new petition in the chancery court of Sumner County. Cates charged the company with contracting to impede "full and free competition" by offering Love free oil to cancel his Evansville contract. The attorney general asked the court to cancel Kentucky Standard's business permit and perpetually enjoin the company and its agents from operating in Tennessee. The chancery court ruled in favor of the state, and the company again appealed to the Tennessee Supreme Court.[18]

On April 11, 1908, Justice M. M. Neil handed down another unanimous decision against Kentucky Standard. The defense primarily argued that Holt had acted on his own when he gave away three hundred barrels of oil. But Justice Neil rejected this proposition, declaring "after fully considering all the circumstances, we are constrained to believe that Mr. Comer sanctioned the acts of Mr. Holt immediately after they were done, if he did not in fact advise them before hand." Moreover, "What was done was within the apparent scope of Mr. Comer's authority, since he was the general manager of the territory."[19] Neil attached particular importance to the fact that the price of Kentucky Standard coal oil rose a cent per gallon "within a week or two" after Evansville Oil left the area and that all price increases originated with the general manager of Kentucky Standard in Cincinnati.[20] The general manager's involvement demonstrated that the top leadership of the corporation was aware of Holt's activities in Gallatin. Justice Neil therefore sustained Kentucky Standard's ouster from Tennessee.

The company then took the case to the United States Supreme

Court. Standard Oil lawyers based their defense on two points: the Tennessee antitrust law unconstitutionally infringed on the congressional commerce power, and it violated the equal protection clause of the Fourteenth Amendment because it denied corporations certain procedural rights granted to individuals—such as preliminary investigation by a grand jury, indictment, jury trial and application of the statute of limitations. On May 2, 1910, Justice Oliver Wendell Holmes handed down a brief opinion for still another unanimous court. Briefly dismissing both objections, he was particularly impatient with the second defense agrument: "The foregoing argument is one of the many attempts to construe the 14th Amendment as introducing a factitious equality [between individuals and corporations] without regard to practical differences that are best met by corresponding differences of treatment."[21] He upheld Tennessee's explusion of Kentucky Standard.

Once again, the legal process seemingly had triumphed over monopoly. The *Literary Digest* noted that several newspapers expressed relief over the ruling. The Philadelphia *Press* declared that the decision "is both interesting and important" since "it calls attention to some of the particular methods of the Standard Oil Company, and upholds a State in treating them as criminal."[22] The San Francisco *Bulletin* remarked that the ruling gave "hope that the power of abstruse technicality is on the wane."[23]

But as before, the state's legal victory was more apparent than real. In 1908 the Jersey Standard board of directors decided to build a refinery at Baton Rouge, Louisiana. In April 1909 they created Standard Oil of Louisiana to manage the new plant and to market petroleum products in the small section of Louisiana east of the Mississippi. Faced with Kentucky Standard's ouster from Tennessee in 1910, Jersey Standard executives simply reassigned marketing in that state to Louisiana Standard before the Supreme Court decision took effect. Thus the Standard combination was able to circumvent the law without losing any business in Tennessee. Company records reveal that in 1910 Louisiana Standard sold 85.5 percent of the kerosene and 81.7 percent of the naptha and gasoline in Tennessee.[24]

In 1904, as Tennessee began its long legal battle with the Standard combination, Kansas became embroiled in a far more extensive oil controversy. Kansas had begun producing oil back in 1891 when a Pennsylvania wildcatter, W.M. Mills, made a modest strike at Neodesha, a small town in the southeast corner of the state. Mills's first well produced only twelve barrels per day, and he ran out of money before he could complete drilling two others. But Mills had faith in Kansas oil, so he traveled to the East in search of financial support. In 1893 he sought a loan from James M. Guffey and John H. Galey of Pittsburgh. Instead of backing Mills, the firm of Guffey and Galey bought his drilling rights. The Pittsburgh oil men then quickly acquired additional leases on Kansas oil land. They invested heavily and drilled about a hundred wells. By early 1895 they had 150,000 barrels of oil in storage, and their wells were producing 1,800 more barrels every day. Yet they did not have an adequate local market for their product, and they could not profitably ship their oil from Neodesha to either Kansas City or Omaha.[25]

In an effort to salvage his investment, Galey met in New York with John D. Archbold and Henry M. Flagler of Standard Oil. In October 1895 they concluded a deal: The Forest Oil Company, a Standard-producing affiliate, acquired Guffey and Galey's Kansas operation for $225,000. Forest Oil then applied the vast resources of the Rockefeller trust to the development of the Kansas-Oklahoma, or mid-continent, field. The company invested over $750,000 in its first two years of operation. It acquired additional oil land, drilled new wells, and constructed pipelines. In 1897 another branch of the Standard combination, the Standard Oil Company of Kansas, completed a refinery at Neodesha with a daily capacity of five hundred barrels.[26]

Some Kansans were apprehensive about the scope and secrecy of the oil trust's activities. In 1896 the Kansas City *Star* complained that "it is practically impossible for anyone to learn accurately the results of the operations of the Forest Oil Company." But most people welcomed the rush of economic activity. The citizens of Neodesha were so pleased with the arrival of the combination that they built a large house in the town's best neighborhood for Forest Oil's first manager. In 1903 the *Star* surveyed the state's recent economic development and observed: "Kansas, struggling a few years ago to pay off a heart-breaking load of indebtedness, is alive today with the hum of factory wheels."[27] Oil was the primary force behind the boom, and prosperity had a remarkably soothing affect on the anxieties of antimonopolists.

The charters of Forest Oil and Kansas Standard soon proved inadequate for the oil combination's projected expansion in the state. Therefore in 1900 the Standard management organized the Prairie Oil & Gas Company with a charter that authorized activity in all phases of the petroleum industry. In 1901 this new company absorbed Forest Oil, while Kansas Standard maintained a separate identity. Prairie Oil & Gas promptly obtained business permits in states throughout the Mississippi valley and erected a vast interstate pipeline and storage network. The dramatic increase in the company's capitalization indicates how rapidly it grew. In 1900 Prairie Oil & Gas had a book value of $300,000; within seven years that figure had risen to $20 million.[28]

The oil trust capped its Kansas organizational structure in 1903 when Standard Oil of Indiana obtained a business permit in the state. Three Standard companies now operated in Kansas: Prairie Oil & Gas produced, purchased, piped, and stored crude oil; Kansas Standard did the combination's refining; and Indiana Standard served as the marketing agent. Although all three companies were legally entitled to engage in most phases of the petroleum business, each carefully remained within its assigned sphere.[29]

In 1901, just as the Spindletop strike was transforming Texas into a major producing state, mid-continent oil production began to soar. Wildcatters discovered several new fields in southeast Kansas, and oil fever seized the region. Over five hundred independent producing companies appeared in a six-month period, most with meager capitalization.[30] Production statistics for the mid-continent field reveal the dimensions of the boom: 189,151 barrels in 1901, 368,848 in 1902, 1,071,125 in 1903, 5,617,527 in 1904, and 12,013,495 in 1905.[31]

The Standard board of trustees welcomed the flood of mid-continent oil because production in the great Appalachian field was declining at an alarming rate. The oil trust therefore expanded its facilities and bought all available crude. Kansas Standard increased the capacity of its refinery at Neodesha to 2,500 barrels per day. Indiana Standard constructed a new 10,000-barrel refinery at Sugar Creek, Missouri, near Kansas City, and laid a 116-mile pipeline from this facility to the Neodesha field. Prairie Oil & Gas erected several complexes of huge storage tanks to hold surplus production. And in 1904 Indiana Standard began work on a 460-mile pipeline from Kansas City to its great central refinery at Whiting, Indiana. In all, the oil trust's Kansas investment reached $15 million by 1905. Still it could not keep up with the ever-increasing flood of oil. In January 1905 Prairie Oil & Gas had over five million barrels of crude in storage, and no faltering in production was in sight.[32]

While rapidly expanding its own facilities, the Standard combination worked to prevent independent producers from finding outside markets. For example, I. E. Knapp, who was the principal independent in the southeast Kansas field, originally shipped his oil by rail to non-Standard dealers in Kansas City and Omaha. But in 1902 the Santa Fe Railroad greatly increased his freight rates. Facing bankruptcy, Knapp gave in and began selling crude to Prairie Oil & Gas in 1903. Standard interests held substantial amounts of Santa Fe stock, and the combination frequently laid pipeline along Santa Fe railroad bed. Hence Kansas producers understandably suspected collusion. Their suspicions increased when the railroad again raised its rates immediately after Indiana Standard completed its Neodesha-Sugar Creek pipeline in 1904. Two separate federal investigations later confirmed that the oil trust and the railroad consistently had manipulated rates to the detriment of Kansas independents.[33]

Like Texas and Tennessee, Kansas had a formidable antimonopoly tradition. In fact, the Kansas legislature had enacted the nation's first antitrust law back in 1889. Still the oil trust brought considerable wealth to the state, and Kansas did not turn against the combination until crude oil prices collapsed in 1904. During that year the value of a barrel of Kansas crude fell from $1.32 to $.72. Prairie Oil & Gas reduced prices even further by shifting to an oil grading policy based on specific gravity. Under this system the company offered $.72 for 32° oil, $.50 for 30° oil, and $.30 for 28° oil at the beginning of 1905. Kansas producers resentfully watched the value of their product decline relative to eastern oil, which usually had a higher specific gravity. They became even more unhappy when they discovered that the price of Standard's kerosene was not going to reflect the drastic drop in the value of crude oil.[34]

The producers began to consider ways to break the oil trust's domination of their industry in the summer of 1904. Orators denounced the combination at mass meetings around the oil region. Some producers advocated the creation of an independent oil company, capitalized at $50 million, that would engage in all phases of the petroleum industry and drive Standard from the state, but this scheme was clearly beyond the producers' means.[35] Samuel M. Porter, a Republican candidate for the state senate in the elections of 1904, urged that Kansas construct and operate a more modest refinery, using prison labor to run the facility. Porter's inspiration for this plan came from the state twine factory at Leavenworth Prison that had been built a few years earlier to compete with the twine trust.[36]

The oil issue emerged at a turning point in Kansas politics. In power since the decline of the Populists, the inept Republican machine had alienated a substantial portion of its own party over the years. In the spring of 1904, as oil prices plummeted, Republican reformers—or boss-busters—completed a successful campaign to drive the party regulars from office. Led by Walter R. Stubbs, the boss-busters captured the Republican state convention and unanimously nominated Edward W. Hoch, a small-town newspaper editor, as their candidate for governor. Hoch responded to the producers' cries for a war on the oil trust by embracing Porter's state refinery plan. Hoch easily defeated his Democratic opponent in November. Samuel Porter won a seat in the state senate in the same election.[37]

The new governor addressed the Kansas legislature the day after he took the oath of office on January 9, 1905:

Our oil interests are in jeopardy. . . . Rather, therefore, than permit the great monopolies to rob us of the benefits of the vast reservoirs of oil which have been stored by the Creator beneath our soil, I am inclined to waive my objection to the socialistic phase of this subject and recommend the establishment of an oil refinery of our own in our own state for the preservation of our wealth and the protection of our people.[38]

With this endorsement, Senator Porter introduced his refinery bill, which called for an appropriation of $410,000 on January 12.[39]

Kansas producers simultaneously began to organize. The Kansas Oil Producers' Association, formed at a statewide conference that convened on January 19, promptly adopted a resolution supporting the state refinery. It also called for bills prohibiting price discrimination, establishing maximum freight rates for oil, and declaring pipelines common carriers. In the following months the association carried out several projects aimed at reducing the power of Standard Oil. Association members helped Senator Porter guide his refinery bill through the legislature. The new organization also started a literary bureau that published anti-Standard pamphlets, collected and reprinted newspaper articles supporting the independent cause, and conducted an extensive direct mail campaign. The Producers' Association financed these efforts through a fifty-cent assessment on every independent oil well in the state.[40]

In New York, the oil barons observed these developments with growing concern and concluded that countermeasures were required. In mid-1904 John D. Archbold instructed the Jennings Advertising Agency, which had helped humiliate Frank Monnett in Ohio, to enhance the oil trust's image on the plains. Malcolm Jennings arrived in Kansas in September, only a few weeks before the state elections. He quickly made contracts with newspapers in about thirty towns to carry articles favorable to the combination. Meanwhile Standard executives in Kansas had joined the public relations campaign, granting interviews to local reporters and writing letters to prominent citizens defending their organization.[41] But these measures proved inadequate to stem the rising tide of opposition.

As 1905 began, the Standard directors dispatched a large number of lobbyists to Topeka to present the combination's case to the Kansas lawmakers. The Oil Producers' Association promptly spread rumors that two bags of gold had arrived by express to be used in a massive bribery attempt. The Kansas speaker of the house appointed an investigatory body—referred to as the "smelling" or "boodle committee"—to look into the matter, but it was unable to pick up any scent. And even Ida Tarbell, on assignment in Kansas for *McClure's,* had to admit that "there was never the least evidence of attempted bribery.''[42] The combination also tried to win public support by dropping the price of refined kerosene four cents per gallon throughout the state, but stubborn critics viewed the decrease as either proof of previous profiteering or a simple market reaction to the drastic decline in crude prices.[43]

On February 10, 1905, the oil trust tried a different approach. Prairie Oil & Gas stopped all work on pipelines and storage tanks, throwing substantial numbers of Kansans out of work. The company then announced that it had ceased purchasing Kansas crude with a specific gravity of less than 30°.[44] This order affected 70 percent of the oil wells in Kansas. A company spokesman explained that "Kansas oil below thirty degrees is a fuel [in contrast to an illuminant] and cannot be refined profitably. The refineries having rejected this oil, the Prairie pipe lines were obliged to cease taking it from the wells.''[45] The spokesman appeared oblivious to the fact that the oil trust also controlled the refineries.

The Standard campaign did not have the desired effect. While the oil combination increased the pressure, Kansas lawmakers rushed through a reform program. In February 1905 the legislature enacted a refinery bill, authorizing the construction of a branch penitentiary and oil refinery at Peru, Kansas, in the heart of the state's petroleum producing region. Despite strong opposition from those who denounced the project as an experiment in socialism, the legislators appropriated the full $410,000 requested by Senator Porter for the project. Three other bills, based on Producers' Association proposals, struck at the oil trust: one prohibited price discrimination between localities in the sale of any commodity, another declared pipelines common carriers subject to regulation, and a third set maximum rates for oil transported by either pipeline or railroad.[46]

On March 2, 1905, Kansas Attorney General C. C. Coleman filed a quo warranto suit in the state supreme court under provisions of the Kansas Antitrust Act of 1889. He charged that Prairie Oil & Gas Company had conspired with several railroads to control the oil transport business and sought to oust the company from the state.[47]

Kansas had launched a broad assault on the oil trust. The state refinery symbolized the legislature's determination to break the combination's stranglehold on the petroleum industry. In fact, the refinery could be little more than a symbol. When completed it would handle only 1,000 barrels per day, while Kansas oil wells produced 27,000 barrels daily. Clearly the state refinery would be incapable of affecting oil prices by itself. Nevertheless, its backers hoped that it would encourage others to build independent installations, thus creating real competition and raising the price of Kansas crude.[48]

The other three bills enacted by the legislature in February 1905 were also far less formidable than they appeared at first glance. The law prohibiting price discrimination between localities restated an earlier provision in the Kansas Antitrust Act.[49] The bills declaring pipelines common carriers and setting maximum pipeline rates were important reforms, but state boundaries severely limited their effectiveness. At the beginning of 1905, the Kansas Standard plant at Neodesha was the only significant oil refinery in the state. Its capacity was 3,500 barrels per day. Therefore, the other 23,500 barrels of Kansas's daily oil production had to be processed at the Indiana Standard refineries in Sugar Creek, Missouri, and Whiting, Indiana. The state of Kansas, of course, had no jurisdiction over interstate rates to those refineries.

Because of the limited effectiveness of the legislative program, the antitrust suit against Prairie Oil & Gas, which operated virtually all petroleum pipelines in the state, was the key to the Kansas anti-Standard campaign. If the state could expel that company, then the petroleum industry would be fundamentally transformed and real competition would be possible. The antitrust suit became even more critical on July 7, 1905, when the Kansas Supreme Court overturned the refinery bill, unanimously ruling that the state oil refinery was a "work of internal improvement" and as such was explicitly prohibited by the state constitution.[50]

While Kansas battled the oil trust, the United States Bureau of Corporations conducted an extensive investigation of the petroleum industry throughout the nation. The federal investigators uncovered mountains of highly damaging information against the Standard combination.[51] Citing the difficulty of obtaining evidence, Kansas Attorney General Coleman withdrew his original petition against Prairie Oil & Gas on March 5, 1906.[52] Using the information supplied by the Bureau of Corporations, Coleman then filed a new quo warranto petition against all three Standard affiliates in the Kansas Supreme Court on October 2, 1906. He charged that the three companies had transferred their stock to Standard Oil of New Jersey. By virtue of its stock ownership, Jersey Standard dominated the entire Kansas oil industry and directed its affiliates to deal exclusively with each other. Thus Prairie Oil & Gas shipped its crude only to Kansas Standard or Indiana Standard refineries, and Kansas Standard sold its refined products only to Indiana Standard for marketing. This restrictive arrangement allowed the combination to control over 75 percent of the retail oil trade and almost all oil pipelines and refineries. Coleman concluded that the oil "producers and the purchasing public have been and are, through the operation of such combinations, secret understandings, pool and monopoly, deprived of full and wholesome competition in said trades and businesses, to the great damage of the people of the state."[53] He therefore called for the ouster of the companies.[54] The court appointed Commissioner L. W. Keplinger to investigate the competitive situation in the Kansas oil industry.

The events in Kansas captured the attention of the national press in 1905. Many of the nation's leading magazines carried feature articles on the Kansas struggle against the oil trust, and newspapers gave the story considerable coverage.[55] The press showered extravagant, premature praise on the prairie trustbusters. William Allen White declared, "Standard Oil with its millions could not prevent the measures aimed at its evil practices from passing the Kansas Legislature by overwhelming majorities."[56] Ida Tarbell appraised the Kansas episode as "a national blow to that arrogant spirit of greed which, for selfish gains, aims to put a bounty of nature in chains; a demonstration that in America men need not, if they will not, be commercial slaves."[57] *The Independent* even noted the spiritual ramifications of the antitrust crusade: "In the churches at Topeka . . . ministers gave thanks for the advantage gained by the people, and asked for divine help throughout the contest."[58]

Before the Kansas Supreme Court struck down the refinery bill, the prospect of prairie socialism also fascinated the press. Editorials violently denounced and stoutly defended what the Cleveland *Plain Dealer* labeled Kansas's "bizarre economic experi-

ments.''[59] Stung by the widespread charges of socialism, Governor
Hoch defended the refinery bill in *The Independent*. He declared
that the bill was a defensive measure, designed to promote competition and private enterprise:

I wish to emphasize and re-emphasize that the State refinery method of
protecting State oil interests is not socialism. It is not the spirit of socialism,
but the very reverse of it. It may have the semblance of socialism, but its
soul is that of competition. Socialism is a heresy which I have studied and
combated for years and the fallacy of which I am more than ever convinced.[60]

Having denied any connection with the socialist ''heresy,'' the Kansas governor abandoned conventional definitions to apply the
dread label to the activities of the oil trust:

No greater question confronts the American people than the control of the
great aggregations of capital, all of them socialistic in character, and which
are antagonistic to the essential element of all national progress, the competitive system.[61]

The press grossly exaggerated the effectiveness of the Kansas antitrust campaign and the significance of the state's socialist experiment. The ephemera of Kansas socialism vanished within six
months. The oil producers, having achieved their legislative goals,
went back to producing oil, and commissioner Keplinger moved
with exceptional lassitude, dragging out his investigation over five
years.

Meanwhile the situation in the Kansas petroleum industry remained essentially unchanged. Mid-continent production increased
dramatically. The region yielded over 22 million barrels in 1906.
Five years later that figure had reached 66 million barrels, a full 26
percent of total national output. But almost all of the new oil was
located in Oklahoma. Kansas production remained static at about 4
million barrels per year, and while major independents such as Gulf
and Texaco were able to enter Oklahoma, only a few minor refiners
challenged Standard in Kansas. In the face of all this new
Oklahoma crude, prices continued to sink. In 1905 Kansas producers had been appalled at the pitiful seventy-two cents that a bar-

rel of their oil commanded. Yet in 1906 the price fell to forty-one cents, and it remained at or below forty cents for several years.[62] The oil trust, as a Kansas economist noted in 1910, could well afford an attitude of "good-natured indulgence" toward the independent oil refiners of Kansas.[63]

Commissioner Keplinger filed his long-delayed report on June 11, 1911. His most important findings confirmed the charge that the Standard combination exercised monopolistic control over the Kansas oil industry.

The Standard Oil Company of New Jersey has in fact whole and complete control over all the companies belonging to the Standard Oil system including the defendants herein, and for all practical purposes is the owner of all properties whereof said companies are the legal owners.[64]

It was the intention of the creators of each of the defendant companies that there should never be any competition between them, and there never has been any competition between them, and there never has been any intention on the part of either of the defendants to engage in any business engaged in or carried out by either of the other defendants.[65]

Not less than 75 per cent of the purchase price and sale of crude oil and of the manufactured products thereof within the country and within the state of Kansas are made by companies controlled by the Standard Oil companies. . . . The prices paid by said companies . . . does practically establish the price for the trade generally.[66]

Incredibly Keplinger ignored his own conclusions and the massive body of facts collected by former Attorney General Coleman and the Bureau of Corporations in order to portray the combination in a favorable light. He noted that competition in the oil industry, although still marginal, had increased. He attributed the drastic decline in crude oil prices to excessive production. Retail prices for refined products also had declined, and the quality had improved. Keplinger ignored evidence of local price cutting by the trust to drive out competition. He professed himself incapable of finding evidence indicating conspiracies with railroads to manipulate freight rates. Moreover, Keplinger emphasized the oil trust's contributions to the state economy: the three Standard companies employed many skilled workers, they invested large sums of money in the state, and they continually developed new technology.[67] Keplinger concluded that the ouster of Indiana Standard would be "a great detriment to the general public."[68] By implication, this judgment also applied to Prairie Oil & Gas and Kansas Standard.

Attorney General Coleman's 1906 quo warranto petition charged the Standard Oil of New Jersey with owning and controlling the three major oil companies in Kansas. It asserted that the three firms each performed a separate function within the industry and avoided competition with the other two, thus depriving Kansas of "full and wholesome competition." Despite its favorable tone, the Keplinger report confirmed these fundamental accusations. The Kansas Antitrust Act of 1889 outlawed combinations "to prevent full and free competition."[69] The Keplinger report therefore provided proof that the Standard companies had been in continuous violation of the law since 1903.

Consequently John S. Dawson, the Kansas attorney general in 1911, had an exceptionally solid case. Recent federal antitrust investigations of Standard Oil, which verified the Jersey Standard ownership of the three Kansas companies, further bolstered his position.[70] Nevertheless Dawson declined to take the matter to court. Instead he negotiated a settlement with the defendant corporations that the state supreme court approved on June 15, 1911.[71]

Dawson offered two reasons for his action. First, he cited Keplinger's conclusion that the corporations neither conspired with railroads to manipulate freight rates nor engaged in other anticompetitive practices.[72] But Dawson had every reason to doubt the accuracy of Keplinger's findings because they conflicted with the conclusions of the federal antitrust investigation and the well-publicized 1906 report of the United States commissioner of corporations. Moreover, the question of anticompetitive business practices was not the determining issue in the case. Dawson's second reason for declining to prosecute was equally feeble. He claimed that recent federal action against Jersey Standard made prosecution unnecesary.[73] Yet federal activity did not remove the combination's liability for antitrust violations in Kansas. Dawson still had an excellent opportunity to expel the Standard corporations and promote competition in the Kansas oil industry, an opportunity he refused to pursue.

Dawson negotiated terms of settlement that raised further questions about his motivation. The companies paid a total of $25,000 in civil penalties, an insignificant sum considering the scale of business interests involved. The penalty was significant, however, as an acknowledgment by all parties of the guilt of the Standard corporations. The settlement decree also imposed several restrictions on the three companies: it prohibited Indiana Standard from producing, transporting, or refining petroleum in Kansas (Indiana Standard refineries were in Missouri and Indiana), it barred Kansas Standard from producing or marketing oil (that company had never engaged in these phases of the industry), and it forbade refining or marketing by Prairie Oil & Gas (again, no alteration of existing practices). In other words, the settlement restricted the charters of the three companies in such a way as to lock them into the anticompetitive structure originally designed by the Jersey Standard board of trustees.[74] The Kansas Supreme Court decree thus guaranteed the perpetuation of the Standard Oil monopoly in the state.

What prompted Dawson to arrange this pro-monopoly settlement? The two most prominent attributes of the Kansas attorney general's public career were respect for the law and faith in the American free enterprise system. A Scottish immigrant, Dawson rapidly advanced in the Kansas legal establishment. He served as assistant attorney general and as state railroad commissioner before becoming Kansas attorney general in 1909. Dawson joined the Kansas Supreme Court in 1914 and served on that body for over thirty years. Because of his commitment to the legal system, he was undoubtedly sincere in sharing the common belief that the federal antitrust effort would adequately deal with the oil trust. In addition, Dawson held no brief against big business. One authority described him as "a republican with progressive tendencies, but [who] confines his activities within the ranks of the party."[75] The American economic system had been good to him, and Dawson apparently had no desire to punish unnecessarily a leading local industry.

Moreover the attorney general served under Governor Walter R. Stubbs, who had led the boss-busters in 1904 and had proven himself a good progressive since his elevation to the statehouse in 1909. He successfully championed state meat inspection, workman's compensation, and campaign contribution disclosure laws. He was a leading supporter of Theodore Roosevelt in 1912.[76] Governor Stubbs viewed the application of sound business practices to government as the key to progressive reform. He had earned over $1 million as a railroad contractor and bank president and had been too preoccupied with business even to vote until he approached the age of fifty. Understandably, he turned to his extensive business experience in an attempt to bring efficiency and economy to Kansas state government. In 1912, Stubbs explained his political philosophy: "I am trying to run a state as I would run a business."[77]

The Keplinger report demonstrated that good businessmen would not drive the Standard companies from the state. The report overflowed with statistics demonstrating the importance of the oil companies to the state economy. It noted that Prairie Oil & Gas operated a vast pipeline system and had 43 million barrels of crude oil in storage. It revealed the extent of the oil trust's domination of refining capacity for Kansas crude: Indiana Standard's capacity was 54,000 barrels per day, Kansas Standard's was 8,000, and Kansas independents had a total of 6,600. The report also noted that Indiana Standard operated 156 tank stations in the state, which employed many people and distributed petroleum products throughout the state.[78] The Standard companies were too vital to the Kansas economy to be expelled, even if the state had to promote, rather than prohibit, monopoly in order to guarantee their continued prosperity.

(4)
The States Strike Out

In November 1904 the voters of Missouri elected Joseph W. Folk, a Democrat, governor of the state. Folk had earned a reputation as a crusader against political corruption in St. Louis, and he intended to dedicate his administration to progressive reform. Herbert D. Hadley won the office of attorney general in the same election. Although a Republican, Hadley fully shared Folk's reform philosophy.[1] The oil controversy, raging across the border in Kansas, made the new attorney general keenly aware of the activities of the Standard combination. In early 1905, as Kansas lawmakers attempted to legislate competition into that state's petroleum industry, Hadley took a hard look at the Missouri oil business. Unlike Kansas, Missouri produced no oil. Nevertheless, three Standard affiliates operated in the state: the Waters-Pierce Oil Company, Standard Oil of Indiana, and the Republic Oil Company. Of these, only Indiana Standard openly acknowledged its connection with the oil trust.[2]

The Standard combination had owned at least 60 percent of the St. Louis-based Waters-Pierce Oil Company since 1878. Although antitrust problems in Texas had forced the reorganization of Waters-Pierce in 1900, the oil trust still secretly held 2,747 of the company's 4,000 shares. In fact, Henry Clay Pierce's unfortunate habit of attracting political and legal controversy prompted the secrecy-loving oil barons to tighten their control substantially over Waters-Pierce in 1904. This move, however, does not mean that the St. Louis firm had ever strayed far from established corporate policy. From the moment of its birth, Waters-Pierce had adhered faithfully to the Standard decree that it restrict its Missouri operations to the southern half of the state.[3]

Standard Oil of Indiana entered Missouri in 1897 as the oil trust's marketing agent in the northern portion of the state.[4] The territorial division between Waters-Pierce and Indiana Standard was so precise that a map of Missouri, with a line tracing the marketing boundary, hung in the offices of both companies to prevent violations. If a Waters-Pierce salesman received an order from the north, he transferred it to Indiana Standard, and Waters-Pierce received similar referrals. The oil trust worked diligently to prevent truly independent marketers from gaining access to the products of its refineries. The huge Indiana Standard plants at Sugar Creek, Missouri, and Whiting, Indiana, sold only to two "independent" retailers in Missouri: Waters-Pierce and Republic Oil.[5]

The Republic Oil Company was formed in New York in 1901, with Jersey Standard holding all of the firm's stock. Republic Oil took over the business of Schofield, Shermer & Teagle, an independent marketing firm that had recently sold out to the oil trust. Company officials took elaborate precautions to prevent disclosure of their connection with Standard Oil. They, like Waters-Pierce, sent all correspondence with their New York headquarters to the back entrance to the Standard Oil Building. The Republic management succeeded in concealing the company's ownership from Missouri officials, the general public, and even most of their own employees for several years. It maintained this elaborate cloak of secrecy because it wanted to continue Schofield, Shermer & Teagle's role as a competitor of both Indiana Standard and Waters-Pierce. By posing as a small, embattled independent, Republic captured the business of customers with a strong antimonopoly bias. In fact, the company's advertisements regularly featured headlines proclaiming "No Trust," "No Monopoly," and "Absolutely Independent."[6]

Together, the three Standard companies controlled 85 to 90 percent of the Missouri oil trade, a level of domination that allowed the combination to set the price for refined products in the state. And the companies used that power to raise their prices at a time when crude oil production was soaring and crude oil prices were plummeting.[7] The average price per gallon of coal oil in St. Louis rose during the period of crude price collapse in the neighboring Kansas oil field: the price per gallon in 1901 was $.073; in 1902, $.080; in 1903, $.092; in 1904, $.093; in 1905, $.095; and in 1906, $.096.[8]

The Standard companies in Missouri routinely engaged in the familiar array of anticompetitive practices. Indiana Standard frequently sold oil from the same tank under a variety of names and at substantially different prices. Republic Oil drastically cut prices in localities where legitimate competition appeared and it granted rebates to retailers who agreed to carry its oil in place of independent products. When the competitors withdrew, prices rose and the rebates stopped. The Standard companies also perfected a system that allowed them to monitor the shipments of independents. Standard agents then took appropriate steps to capture the independents' market.[9]

Attorney General Hadley uncovered the first evidence of Standard Oil's conspiracy in restraint of trade during a freight rate investigation in early 1905. He found that Indiana Standard did no business in St. Louis and that Waters-Pierce remained completely out of the Kansas City market. He was certain that this lack of competition in Missouri's two leading markets was more than a coincidence. Realizing the tremendous political advantage in a successful prosecution of the hated oil trust, the attorney general quietly began to gather information. Independent oil dealers and former employees of the Standard companies willingly provided evidence.[10] And on March 29, 1905, Hadley filed a quo warranto petition in the Missouri Supreme Court, charging that Waters-Pierce, Indiana Standard, and Republic Oil had formed an illegal combination between 1901 and 1905 to prevent competition and control retail prices. Furthermore, the companies had deceived the public by posing as independent and competing concerns. The attorney general asked that the court revoke the Waters-Pierce charter and cancel the Indiana Standard and Republic Oil business permits.[11]

In May 1905 the Missouri Supreme Court appointed a special commissioner, Robert J. Anthony, to take testimony in the case. Hadley promptly began an investigation that lasted two years. The

Missouri attorney general questioned over a hundred witnesses in several cities, including St. Louis, Kansas City, Cleveland, and New York. All the leading figures in the Standard hierarchy, with the notable exception of John D. Rockefeller, eventually appeared. And despite the determined opposition of a small army of Standard Oil lawyers, Hadley compelled the three companies to produce their books for examination.[12]

The Missouri inquisitor captured the attention of the national press during his first session with the oil barons in New York City. Hadley secured about forty subpoenas from the New York Supreme Court in late 1905 and received a flood of favorable publicity as reporters followed the chaotic attempts of the Standard magnates to avoid the process servers. The *Arena* reported that Charles M. Pratt, a member of the Standard board, "cancelled his extensive social programme for the winter and hastily fled from Brooklyn."[13] Walter Jennings, another board member, narrowly eluded the grasp of Hadley's agents: "At midnight, in a hired boat, he fled across the Sound, and since has been practically in exile in Fairfield, Connecticut."[14] According to the *New York American*, John D. Rockefeller remained on one of his estates, safely beyond the reach of the law:

John D. Rockefeller, much as he likes to spend Christmas in the city, decided this year that the Pocantico Hills were more comfortable. There, in the heart of his domain, surrounded by detectives and with pickets on guard before every approach, he has been a prisoner since November 1st.

Time and again process-servers in various disguises have succeeded in passing the the pickets, but never have they penetrated beyond the inner guard of detectives. When discovered they have been handled roughly and promptly ejected by the oil king's minions.[15]

Hadley's agents, however, usually overcame such resistance. In an article in the St. Louis *Post-Dispatch*, M. E. Palemdo, a process server, proudly explained how he had cornered Henry H. Rogers. Palemdo discreetly positioned himself near Rogers's opulent New York townhouse. Shadowed by a bodyguard, Rogers moved rapidly one morning from the front door of his home to the back seat of his car. As the chauffeur sped off, Palemdo sprung from his hiding place to the running board of Rogers's car and inquired, "Is this Mr. H. H. Rogers?" The startled tycoon did not reply. Undaunted, Palemdo threw the subpoena at Rogers, flashed the court order, and jumped from the car.[16]

When Hadley's hearings opened on January 5, 1906, a formidable array of lawyers were present to defend the Standard Oil leaders. They did everything in their power to impede the young, inexperienced attorney general from Missouri. On the first day, the Standard lawyers invoked an old New York statute that required all testimony to be taken down in longhand. After considerable argument, they conceded that a typewriter could be used, but they adamantly refused to allow shorthand.[17] The obstructionism peaked during Henry H. Rogers's testimony. The oil baron opened by objecting to the presence of newspaper artists and smokers. He then refused to answer most questions. At one point, he even claimed ignorance about the location of the oil trust's main office. When Hadley asked who owned the three companies operating in Missouri, Rogers refused to answer "on advice of counsel." Although he temporarily preserved the combination's secrets, Rogers's arrogance provoked considerable hostility in the press. Even normally conservative papers denounced his behavior.[18]

Earlier in the investigation, Hadley had demanded that Republic Oil turn over its stock book to establish the ownership of the corporation, but the company secretary had refused, claiming that stock ownership was immaterial to the investigation. The question then went to the Missouri Supreme Court.[19] Following Rogers's refusal to testify on this same matter, Hadley asked the New York Supreme Court to require Rogers to show cause why he did not respond. The New York court, however, declined to take action before the Missouri court had reached its decision.[20] Hadley therefore returned home to await the Missouri court's decision on the relevance of stock ownership. On February 26, 1906, that court ruled in Hadley's favor.[21] Early in March, Standard attorneys informed Hadley that the combination would be more cooperative.

On March 24, Hadley was back in New York and had Henry H. Rogers on the stand. He quickly went to the main point in the case: "I ask you again whether all or a majority of the stock of the Standard Oil Company, of Indiana; the Republic Oil Company and the

Waters-Pierce Oil Company is owned or controlled, directly or indirectly, by the Standard Oil Company, of New Jersey.''[22] As Rogers sat in defiant silence, a Standard lawyer read the following statement:

Subject to the objection that it is immaterial and irrelevant, it is admitted for the purposes of this case only, that now and during the period covered by the information, the majority of the stock of the Standard Oil Company of Indiana and the stock of the Republic Oil Company is held for the Standard Oil Company of New Jersey, and that all stock of the Waters-Pierce Oil Company on the books of the company in the name of M. M. Van Buren is held for the Standard Oil Company of New Jersey.[23]

Hadley had established the key issue in the case. The New York *World* headline for March 25, 1906, exclaimed: "Rogers Beaten, Oil Trust Owns Up to Monopoly."[24]

Hadley still wanted to examine Henry Clay Pierce concerning Waters-Pierce operations in Missouri. The attorney general's agents had nearly cornered Pierce in January 1906. On the eleventh of that month, the *World* reported that "the best subpoena sleuths in New York were started out yesterday after H. Clay Pierce. . . . Subpoena-servers corralled him [at the Waldorf-Astoria] last week, but by locking himself in a bathroom he escaped service until the intruders were driven away."[25] He reportedly then fled to the safety of his yacht. But Pierce grew tired of evading detectives and finally agreed to testify in St. Louis in September 1906. On the stand, Pierce admitted that Jersey Standard held a controlling interest in the Waters-Pierce Oil Company. He also acknowledged that his firm and Indiana Standard had divided Missouri into two separate marketing territories in order to avoid competition. Pierce placed responsibility for all illegal activities in Missouri squarely on the board of trustees in New York.[26]

Hadley concluded his investigation, the most comprehensive analysis of the Standard combination to date, in January 1907. Commissioner Anthony's report, submitted to the Missouri Supreme Court on May 24, 1907, contained mountains of damaging material. The report documented Standard Oil of New Jersey's ownership of Waters-Pierce and Republic Oil, despite the fact that both companies posed as independents. It proved that Waters-Pierce and Indiana Standard operated in separate marketing zones. It also detailed repeated instances of price cutting, rebates, and espionage. Commissioner Anthony concluded that the defendant corporations had fixed prices and limited trade in petroleum products, while they "deceived and misled the public into the belief that they were separate and distinct corporations pursuing an independent business." These activities were "all contrary to and in violation of the laws and public policy of the state of Missouri."[27] Hadley's triumphant investigation produced immediate political results; he was elected governor of Missouri in 1908.[28]

On December 23, 1908, nineteen months after Commissioner Anthony had submitted his report, the Missouri Supreme Court found all three companies guilty. Speaking for a five to two majority, Justice Archelaus Woodson turned aside the familiar claims by Standard Oil lawyers that the Missouri antitrust law violated the due process and contract clauses of the federal constitution and infringed upon the congressional power over interstate commerce. He noted that other state antitrust statutes had been upheld on numerous occasions in the face of identical objections.[29]

Justice Woodson observed that the massive record presented by the prosecution contained overwhelming proof of the combination's guilt. That record abounded in instances where the Standard organization had "withered the energies of the competitors; blighted individual investments in legitimate business; [and] driven small and honest dealers out of business for themselves."[30] The judge foresaw grave dangers in allowing Standard Oil to maintain its monopoly. If such flagrant abuses continued unchecked, "It would be only a question of time until they would sap the strength and patriotism from the very foundations of our government, overturn the republic, destroy our free institutions, and substitute in lieu thereof some other form of government."[31] Woodson therefore canceled the business licenses of Indiana Standard and Republic Oil and revoked the Waters-Pierce charter, fined each company $50,000, and gave Indiana Standard and Republic Oil until March 1, 1909, to terminate their operations in Missouri.[32]

While taking a passionate stand against the abuses of the Standard combination, Woodson demonstrated a keen sensitivity to the economic consequences of his decision:

In arriving at the conclusions stated, we have not lost sight of the vast financial interest involved, nor the magnitude of the business institutions to be affected thereby. Nor are we unmindful of the fact that an assault made upon those vast interests is also a serious wound inflicted upon the material welfare of the state itself, by banishing so much capital therefrom, and otherwise disturbing the financial and business interests of the country.[33]

Because of these considerations and to safeguard the rights of Waters-Pierce minority stockholders, Woodson offered Waters-Pierce a way to save its charter: if the company paid its fine and immediately severed its ties with the oil trust, it could continue to do business in the state.[34]

Dissenting opinions by Justices Walter W. Graves and Henry Lamm revealed even greater concern for property rights and the state's economy. Graves thought that the court treated Waters-Pierce unfairly because it proposed to revoke that company's charter, while it merely sought to oust the other two defendant corporations from the state. Thus Waters-Pierce received unequal punishment because it had the misfortune of being chartered in Missouri. Moreover, by revoking this charter, the court subjected the minority Waters-Pierce stockholder—Henry Clay Pierce—to harsher treatment than the real culprits—the Jersey Standard board of trustees in New York.[35] Justice Lamm, on the other hand, worried about the economic impact of expelling Indiana Standard: "That corporation had vested interests in the State in a plant of great value and a great business in the useful line of refining oil."[36] Lamm therefore wanted to suspend the ouster, dependent only upon Indiana Standard's "good behavior." Both Graves and Lamm admitted that all three companies were guilty of repeated antitrust violations, but their concern for the state's economic welfare and their respect for vested property rights led them to back away from strict enforcement of the antitrust law.

A few months later the Missouri court gave in completely on the Waters-Pierce issue. On March 9, 1909, the court suspended its revocation of that company's charter, noting that Waters-Pierce had paid its fine and had "given satisfactory evidence" of its willingness to obey state law.[37] This ruling was a triumph for Judge Graves. In his earlier dissent, Graves had emphasized the critical importance of treating local business interests fairly. The revocation of the Waters-Pierce charter, he had argued, would have destroyed an important St. Louis-based corporation that marketed products throughout the Southwest. Furthermore it would have severely damaged a leading local investor, yet it would have only indirectly punished the oil trust in New York. The state of Missouri could not afford to treat its business community in such a fashion.

Justice Woodson, who wrote the original decision in the case, exposed the majority's capitulation to Waters-Pierce in a forceful dissenting opinion. What was the "satisfactory evidence" that the company had submitted to prove its withdrawal from the oil trust? It was the following resolution, adopted by the Waters-Pierce board of directors on February 13, 1909:

Resolved that this company, protesting that it has never consciously or knowingly violated any of the provisions of the laws of this State, nevertheless, does hereby accept the terms and conditions of the order or decree of the Supreme Court of Missouri, entered in the cause of the State upon the information of Herbert S. Hadley, Attorney-General, against the Standard Oil Company of Indiana, the Republic Oil Company of New York, and Waters-Pierce Oil Company, and does hereby express its willingness to abide by the same.[38]

Jersey Standard still owned 68 percent of Waters-Pierce stock, and the resolution did not express any intention of altering the stock control of the company. The firm did not even admit to the years of continuous antitrust violations, so recently documented by the Hadley investigation. Woodson asked his colleagues: "Can any intelligent, fair-minded, disinterested person believe for a moment that these respondents have severed their trust relationships with each other or with the Standard Oil Company of New Jersey? We think not."[39] Justice Lamm joined Woodson in dissent, but the majority did not see the case that way. They were looking for a way to allow a substantial business to remain in the state and seized upon the feeble Waters-Pierce resolution as proof of the company's withdrawal from the oil trust.

Indiana Standard and Republic Oil, both out-of-state corporations, were initially in a less favorable position than Waters-Pierce.
Standard lawyers began their efforts on behalf of those firms by filing a petition for rehearing on the business permit revocations.
Then, on February 2, 1909, Indiana Standard lawyer Frank Hagerman made a startling proposal to the Missouri court: the oil trust
offered to form a new corporation that would absorb the Missouri
property of both Indiana Standard and Waters-Pierce. The stock in
the new firm would be placed in the hands of two trustees for a
four-year period; the state would select one trustee, and the oil trust
would choose the other. These trustees would oversee the affairs of
the company to ensure full compliance with the law. In the event of
a policy dispute, the Missouri Supreme Court could intervene.
Through this unusual plan, which originated at Standard
headquarters in New York, the oil trust sought to save Indiana
Standard's Sugar Creek refinery. As Hagerman explained,

Next to that at Whiting, Ind., the Sugar Creek Refinery at Kansas City is
the largest in the United States. Since it was opened the fuel-oil industry has
grown to such proportions that the sudden shutting off of the supply would
mean almost incalculable loss and confusion for a very large number of
enterprises.[40]

This scheme would have enabled the oil trust to maintain ownership—if not complete, immediate control—of the new Missouri
corporation. Moreover, the new company would have integrated
the Missouri marketing operations of Indiana Standard and
Waters-Pierce. Hadley's well-publicized hearings had rendered obsolete the old system of marketing territories. The Standard leadership was willing to abandon traditional anticompetitive business
practices to adjust to the new climate of opinion. Hagerman's proposal showed that the oil trust was not afraid of state regulation.
Years of experience had convinced the Standard directors that state
officials were extremely cautious when dealing with important industries. In fact, the oil barons were ready and willing to use state
regulation to maintain their dominance of the petroleum industry.

The Missouri Supreme Court, however, denied motions by Indiana Standard and Republic Oil for a rehearing and revoked the
business permits of both companies.[41] It ignored Hagerman's pro-

posal. Nevertheless Indiana Standard remained in business in Missouri pending appeal to the United States Supreme Court. Republic Oil had been absorbed by Indiana Standard in March 1906. The Hadley investigation had destroyed Republic's pose as a small independent. Hence the Standard combination had no reason to continue operating the company.

Over three years later, on April 1, 1912, the United States Supreme Court ruled on Indiana Standard and Republic Oil's appeal. In a brief, unanimous opinion, Justice Lucius Q. C. Lamar rejected defense contentions that the companies had been deprived of their property in violation of the due process and equal protection clauses of the Fourteenth Amendment. He upheld the Missouri court's expulsion of both companies from the state.[42]

But Indiana Standard continued to fight. On May 1, 1912, Standard attorneys asked the Missouri Supreme Court to allow the company to remain in the state.[43] The lawyers noted that Indiana Standard had property valued at $3.75 million in Missouri. The company supplied a substantial portion of the state's petroleum products and employed large numbers of Missourians. Therefore the interests of the state would not be served by outright expulsion.[44] The Indiana Standard management took another step at this time to guarantee that the judges understood the possible consequences of strict antitrust enforcement: it suspended all construction at the Sugar Creek refinery and began laying off workers.[45]

But the Missouri Supreme Court refused to back down. Declaring that the time limit for modification of the judgment had expired, it stood by its expulsion order on February 12, 1913.[46] Indiana Standard immediately started shipping crude oil to its refinery at Wood River, Illinois, and workers made final preparations to close down the Sugar Creek operation. The Kansas City Commercial Club started a drive for legislation that would allow Indiana Standard to remain in the state. Pressure from the business community, which feared the consequences of the plant closing, soon induced the state legislature to pass a bill to prevent such an occurrence. The new law provided that any out-of-state corporation with a factory in the state and subject to an ouster order could remain in business on the sole condition that it pay three times the normal fee for a new business permit.[47]

On April 9, 1913, Missouri Governor Elliot W. Major vetoed the Indiana Standard bill. Major had served as state attorney general during the preceding Hadley administration, and he had argued the Indiana Standard case before the United States Supreme Court. He labeled the bill an example of legislative recall and declared that the "laws must not be nullified merely to relieve a particular case."[48] Although the governor acknowledged that the ouster of Indiana Standard might cause economic disruption, he noted that the Sugar Creek refinery could be sold to another oil company. Even if strict antitrust enforcement meant serious hardship, "the laws should not be paralyzed or destroyed for that reason."[49]

The governor's veto dumped the Indiana Standard controversy back in the lap of the Missouri Supreme Court. In February 1913, a lawyer acting on behalf of the citizens of Sugar Creek again had asked the court for a rehearing. On May 10 the court bowed to the mounting pressure and appointed a commissioner to reexamine the matter.[50] Indiana Standard representatives emphasized two points during the hearings: they claimed total independence from the oil trust, and they emphasized the importance of their business to the state. The company based its claim of independence on the United States Supreme Court's dissolution of the Jersey Standard holding company in 1911. But this decision had not altered the fact that a small, cohesive group of New York City oil barons still held a majority of stock in all the Standard companies. Moreover the recent Supreme Court decision had no bearing on the company's culpability for previous crimes. On the other hand, Indiana Standard's place in the state economy was substantial and undeniable. During the hearing, company spokesmen revealed that their firm planned $1 million in new investments in Missouri.[51]

On July 28, 1913, the court, in a per curiam decision, suspended the writ of ouster on the sole condition that Indiana Standard obey the law in the future. The court did not bother to explain the reasoning behind this action or attempt to reconcile this ruling with its previous statement that the time limit for modification of judgment had expired.[52]

Justice John C. Brown submitted a lone dissent. He noted the court's unjustified reversal of position; he argued that the court had in effect granted Indiana Standard pardon because the time limit for modification of judgment had lapsed; and he asserted that only the governor had the authority to grant pardons.[53] Moreover, Brown objected to the court's action as a grossly unequal application of the law:

The defendant stands convicted of having wrung from the people of Missouri a million dollars by criminal methods, and if I possessed the power to pardon it (which I do not) I could not obtain the consent of my own conscience to do so. It shocks my idea of justice to see small criminals incarcerated in the penitentiary, while others, equally as guilty, are granted clemency.[54]

Since the court did not explain its decision, its motives for suspending the writ of ouster against Indiana Standard are unclear, but the judges were undoubtedly influenced by the intense pressure applied by the Sugar Creek refinery workers, the Kansas City Commercial Club, and the oil company itself. The court also was aware that the state legislature was on record as strongly supporting the readmission of Indiana Standard. Even Herbert Hadley, who had been elected governor of Missouri in 1908 on the strength of his oil trust investigation, thought that the company should remain in the state in order to save jobs.[55]

The decision also reflected the majority's concern for vested property rights and economic development. Henry Lamm, who was now chief justice of the Missouri court, had emphasized the size of Indiana Standard's investment and its value to the state economy in his dissent from the original Standard Oil decision of 1908. Justice Woodson, who had written the majority opinion in that case, had expressed a similar awareness of the need to cultivate large-scale enterprise. The pressure campaign merely reinforced the pro-business attitude already present on the bench. The court backed away from the writ of ouster and worked out a settlement with Indiana Standard—just as it earlier had worked out a settlement with Waters-Pierce—because it desired to promote industrial development.

In addition to the court fights in Ohio, Texas, Tennessee, Kansas, and Missouri, the Standard Oil legal staff was fending off antitrust prosecutors in five other states and the Oklahoma territory. In all, thirty-three separate suits were filed against the combination between 1890 and 1911. Yet this additional litigation was equally unsuccessful.[56] A Nebraska effort to expel Standard Oil of Indiana on the grounds that the company was a member of the trust failed because of the difficulty of obtaining evidence. Minnesota and Iowa both unsuccessfully tried to oust that same firm for illegal price discrimination. Arkansas filed suit against the Waters-Pierce Oil Company for Henry Clay Pierce's long-standing affiliation with the Standard group but never took its case to court. Oklahoma also moved against Waters-Pierce, only to compromise its suit when the company promised to build a $150,000 refinery in the area. Three separate prosecutors in Mississippi filed against Standard Oil of Kentucky, one calling for a fine of $1.4 million. These efforts, however, all vanished from public view without discernible effect.[57]

Several reasons stand out for Standard Oil's spectacularly successful evasion of the state antitrust laws, but the combination's vast wealth was the overriding factor. This wealth commanded the services of the nation's greatest lawyers, like Joseph H. Choate, whose subtle distortion of Ohio Attorney General David Watson's quo warranto petition before the state supreme court in 1891 prevented the loss of Ohio Standard's charter. Backed by virtually unlimited financial resources, oil trust lawyers could exploit the drawn-out process of appeal to the limit. Tennessee expelled Standard Oil of Kentucky in 1907, but three years passed as Standard lawyers, losing repeatedly, took the case from courthouse to courthouse, eventually presenting the dispute to the country's highest tribunal. While their legal staff provided cover, the oil barons reassigned marketing operations in Tennessee to another subsidiary corporation. When the United States Supreme Court finally upheld Kentucky Standard's expulsion from Tennessee in 1910, the combination was able to continue its operation in that state without the slightest inconvenience.

Standard Oil also had the money to hire the services of powerful allies in the course of an antitrust suit. When independent oil producers pressured the Kansas state government to restrict Standard operations in 1905, the oil barons dispatched lobbyists to Topeka to work on the state legislature. Standard directors paid the Jennings Advertising Agency to run advertisements, disguised as news items, that were damaging to prosecutors in both Ohio and Kansas. Most important, Standard money bought politicians. Ohio Standard officials hired Senator Joseph A. Foraker to "advise" its legal staff on the Ohio antitrust suits of 1898 and to pressure crusading Ohio Attorney General Frank Monnett into calling off the whole matter. Henry Clay Pierce paid Congressman Joe Bailey to arrange the Waters-Pierce Oil Company's readmittance to Texas in 1900 and to head off subsequent antitrust challenges.

Standard Oil's massive contributions to the Republican party also gave the combination political influence.[58] GOP leaders in Ohio worked hard on behalf of their major contributor. In 1892 Marcus Hanna bluntly reminded Attorney General David Watson of the value of good relations with Standard Oil and then urged him to drop his suit against the oil trust. The party denied Frank Monnett renomination for the office of attorney general after he began several suits against Standard. It even blocked Monnett's brother-in-law's nomination for that same post in 1902. Standard officials felt free to advise leading figures in the Ohio Republican party on candidates for the state judiciary. In 1902 Standard director John D. Archbold wrote to Joseph Foraker urging the reelection of Jacob F. Burket to the Ohio Supreme Court. According to Archbold, the Standard board felt "very strongly that his [Burket's] eminent qualifications and great integrity entitled him to this further recognition."[59] Burket had previously played a key role in Standard's escape from the affects of the Ohio antitrust law.

The oil trust flexed its economic muscle in less blatant but no less effective ways. The petroleum industry was the mainstay of many local economies around the country, and Standard officials reminded the appropriate politicians of that fact when faced with an antitrust threat. In 1904 when Kansas began a reform program that included an antitrust suit, Prairie Oil & Gas, a Standard pipeline affiliate, halted all construction in the state and threw substantial numbers of Kansans out of work. At the same time, the company stopped buying Kansas crude with a specific gravity of less than 30°, a policy affecting 79 percent of the wells in the state. Indiana Standard laid off some workers and began making plans to close its Sugar Creek refinery near St. Louis while the Missouri Supreme Court reconsidered an order expelling that company. The combination also granted favors. Indiana Standard lowered kerosene prices in Kansas in 1904 in an attempt to dampen the momentum of the independent oil producers' antimonopoly crusade. Waters-Pierce promised to build a $150,000 refinery in Oklahoma and thus paved the way for the compromise of an ouster suit.[60] The prospect of a $1.5 million Indiana Standard investment in Missouri helped convince the state supreme court to rescind its ouster order against that firm.

The attitude of prominent members of the legal establishments in various states toward private property and industrial development was yet another reason for the consistent failure of these suits. In certain comparatively rare instances, judges and prosecutors exhibited overt favoritism toward the Standard organization. In 1900 Judge Jacob Burket joined two of his colleagues in voting not to hold Ohio Standard in contempt for failing to carry out the dissolution directives of 1890. The judges ruled in favor of the trust despite proof that the dissolution of 1892 was a sham. Two attorneys general of Ohio displayed a similar allegiance to the combination. John Sheets dropped Monnett's antitrust suits in 1900 without explanation and despite the mass of evidence that already had been gathered in these cases. For the first three years of his term as attorney general, Ulysses Denman stubbornly refused to pursue the antitrust cases against Standard Oil that he had inherited from his predecessor. He then dismissed all charges against the trust.

This type of behavior, however, was the exception. Around the turn of the century most American lawyers and judges still felt the traditional antipathy toward trusts, and they clung to common-law restraints upon monopoly. They usually rejected the arguments of corporation lawyers that they follow the British courts in relaxing restrictions upon big business. Nevertheless virtually everyone associated with these cases had profound respect for property rights. This attitude frequently appeared in judicial decisions forcefully condemning the Standard organization, and it tempered judgment against the combination on several occasions. A good example of the power of this attitude is the Missouri Supreme Court's treatment of Waters-Pierce. In 1908 the court convicted that company of blatant antitrust violations and revoked its charter. But because of the judges' concern for the property rights of the notorious Henry Clay Pierce, who owned 40 percent of the company, they allowed Waters-Pierce three months to sever its ties with the trust. The court then accepted an unsubstantiated statement by the company as proof of this severance and allowed Waters-Pierce to remain in the state.

The almost universal passion among state officials for industrial development was an even more significant factor in these cases. This desire determined the outcome of the Kansas suit against the combination. Commissioner Keplinger, who was appointed by the Kansas court to examine the oil industry as a result of this case, submitted a report that, in addition to demonstrating the trust's guilt, underscored the importance of the Standard group to the state economy. And Attorney General John Dawson, after reviewing the economic facts, promptly negotiated a settlement that left the Standard empire intact in Kansas. Perhaps the ultimate expression of state politicians' desire to promote local industrial development was the Texas legislature's behavior in the Bailey investigations of 1900 and 1906. Texas had a long tradition of extravagant antimonopoly rhetoric, and Waters-Pierce was one of the favorite targets of would-be trustbusters. The investigations into Bailey's dealings with Henry Clay Pierce made it apparent to all that the Texas lawmaker had violated sound ethics. But in the end the Texas political establishment was unwilling to punish one of its own for being too friendly to big business. Texas politicians clearly understood that such action was not the way to promote economic growth.

The American federal system also impeded state officials and contributed to Standard's success. Antitrust investigations were always arduous, but it was doubly hard for state prosecutors to extract information from out-of-state corporations. In some instances this problem simply made the investigation time-consuming and difficult. Frank Monnett traveled repeatedly from Ohio to New York City to question both John D. Rockefeller and John D. Archbold but received only minimal information for his efforts. Similarly Missouri Attorney General Herbert Hadley went to New

York only to meet stiff resistance on the witness stand. In other cases investigative problems posed by the federal system prevented suits from ever getting off the ground. In 1899 Nebraska trustbusters set out to prove the obvious proposition that Standard Oil of Indiana was part of the Rockefeller organization. But fact-finding forays to both Chicago and New York failed to uncover the needed evidence, and in 1901 the matter was dropped.[61]

The federal system also created complexities that gave the combination room for legal maneuver and, if necessary, corporate reorganization. Following the 1892 Ohio Supreme Court directive that Standard Oil of Ohio end its ties to the combination, the Standard board, safe in its New York headquarters, devised a scheme for perpetual dissolution that successfully held the organization together for seven years. When faced with another antitrust challenge from Ohio in 1899, the combination took on an entirely new organizational structure. The board reshaped Standard interests into a holding company under the lenient laws of the state of New Jersey. Despite its new corporate form, the leadership, remained intact and business operations continued as before. Out-of-state reorganizations also proved to be the salvation of the Waters-Pierce operation in Texas. Devised under a cloak of secrecy in New York, the Waters-Pierce reorganization in 1900 provided that the majority of Waters-Pierce stock be secretly returned to the trust as Waters-Pierce returned to Texas.

Finally, the trust emerged victorious because of fundamental inadequacies in the opposition. In Ohio some foes of Standard Oil were simply inept. George Rice committed a major blunder in 1899 when he told the press that Standard had attempted to bribe Attorney General Monnett. The attorney general himself promptly compounded the problem by taking the unsubstantiated bribery charge to court. Only a year earlier Monnett had also charged, without conclusive evidence, that Ohio Standard officials had burned corporate books pertinent to his antitrust investigation. Both of these unfortunate episodes damaged the attorney general's credibility. In Texas prosecutors exhibited different, but no less serious, failings. Governor Hogg and Attorney General Culberson displayed incompetence in 1894 when they charged the New York leadership of Standard Oil with criminal conspiracy in Texas because the law required actual physical presence in the state at the time of an alleged crime to justify extradition. Governor Sayers and Attorney General Smith in 1900, as well as Attorney General Davidson in 1906, all made effective antitrust enforcement impossible because of their overriding concern with political advantage.

Given the underlying economic realities, the attitudes of most lawyers and judges about private property and industrial development, the procedural problems, the limits of state legal authority, and the frequent ineptitude of state prosecutors, these suits had little chance of success. The question that then arises is why there were so many antitrust suits against Standard Oil. The major outbreak of litigation occurred around 1905, long after results of the early experiences in Ohio and Texas were evident to all knowledgeable observers. Still there was strong public demand for action of any kind against the oil trust, and there were potentially great political rewards for politicians who took on the oil monopoly. Furthermore the inevitable collapse of an antitrust suit was always far less visible than the well-publicized opening moves. Thus prosecutors had much to gain and very little to lose by taking the oil trust to court. This litigation proved that a state government was no match for a major national corporation. The remaining question was whether the national government, with its broader reach and greater resources, could enforce its antitrust law any better than the states could enforce theirs.

(5)
The United States versus Standard Oil

By 1900 Standard Oil had already established itself as the prototype of the dominant form of capitalist enterprise in the twentieth century: the large multinational corporation. From their New York City headquarters, the oil barons directed an empire that spanned North America and reached deep into Europe and Asia. Rockefeller and his colleagues had been battling Russian producers for customers in Great Britain and on the Continent since the 1880s. Standard agents began cultivating markets in exotic regions—including Siam, Borneo, and China—in the following decade. Although he remained the bête noire of antimonopolists everywhere, John D. Rockefeller had eased himself out of the day-to-day management of Standard Oil by 1897. John Archbold, a jovial Irishman and self-confessed "clamorer for dividends," then assumed direct control of the organization. Other members of the Standard Oil group, such as Henry Rogers, William Rockefeller, Henry Flagler, and Charles Pratt, continued to provide potent managerial talent, and Rockefeller himself kept in daily contact with headquarters through a direct wire from his estate.[1]

In 1899 the Standard Oil Company of New Jersey became the corporate device through which the oil barons avoided Frank Monnett's Ohio antitrust crusade. That year the directors increased the number of Jersey Standard shares from 100,000 to 1,100,000; the corresponding capitalization rose from $10 million to $110 million. They then authorized the exchange of old trust certificates and shares in various Standard constituent companies for this new stock. The transfers proceeded rapidly, and Standard Oil of New Jersey soon became the repository for the stock of the entire organization. The federal government estimated that Jersey Standard directly held a controlling interest in sixty-five companies and indirectly controlled—through stock ownership by subcompanies—forty-nine others. Only ten men, or their estates, possessed more than half the stock in this intricate construct of 114 corporations. John D. Rockefeller alone held more than a quarter of all Jersey Standard stock, and his brother William owned another 11,700 shares.[2]

The empire of Standard Oil did not depend on drilling oil wells. The apogee of Standard's control over national crude production was the 33 percent share it achieved in 1898, but the rapid emergence of prolific new fields soon cut deep into that figure. The Appalachian field, centered in western Pennsylvania and the source of some of the highest-quality crude, was the birthplace of the industry. The Lima-Indiana field, located on the Ohio-Indiana border, began major production in the late 1880s. Between 1895 and 1900 these two regions alone produced 95 percent of the nation's oil supply. The discovery of oil along the Gulf Coast, symbolized by the great Spindletop strike of 1901, marked a major shift in the geography of oil production. Within the next five years the mid-continent field, located in Kansas and Oklahoma, the California field, and the Illinois field all started to yield oil in quantity. By 1906 these four new areas were producing almost two-thirds of the nation's crude, and Standard's share of total production had fallen to 11 percent.[3]

Despite its diminishing control of the wells, Standard was able to dominate the producers through control of the pipelines; by the turn of the century, they provided the most efficient method of transporting crude oil to the refineries. Oil was then a relatively cheap commodity; hence transportation expense was one of the most important components in the total cost of production. In 1904 the trust operated a 40,000 mile pipeline system capable of transporting crude from the Oklahoma territory to the Atlantic. Through this system the oil barons effectively controlled 85 to 95 percent of production in the Appalachian, Lima-Indiana, Illinois, and mid-continent fields. Although Standard had far less control over the booming new fields on the Gulf Coast and in California, the federal Bureau of Corporations concluded that the oil trust owned so much of the national pipeline system that it dominated oil production throughout the United States.[4]

The oil trust achieved supremacy in pipeline transportation through a variety of tactics. Standard officials bought land or secured rights of way across proposed routes of potential competitors. They enlisted the support of railroads in blocking new lines. They even established bogus competitors who handled the business of oil producers with unusually strong antimonopoly feelings. But the most effective method of weakening an independent pipeline was the payment of premiums to that line's customers. Standard officials offered oil producers using a competing line five to ten cents per barrel above the market price to induce them to switch to the Standard network. The oil trust worked diligently to prevent independent producers and refiners from sharing the benefits of their hard-won pipeline monopoly. Standard refused to supply oil to refiners outside the combination, declined to provide service to points essential to the operation of independents, or set unreasonable regulations as to the quantity of crude required to qualify for shipment. The combination also charged excessive rates on the independent oil that it agreed to transport.[5]

The Standard group possessed a nationwide complex of refineries to match its pipelines. The transportation network allowed the oil barons to locate their refineries near major markets or the main channels of national and international trade, while independents had to build their plants near the oil fields and then ship their products to market. The combination operated twenty-three refineries in 1904. There were five Standard plants in the New York area alone, including the nation's largest at Bayonne, New Jersey. The Standard Philadelphia refinery was the second largest in the United States, and a huge installation at Whiting, Indiana, served the major midwestern markets. The independents operated a total of 75 refineries at that time, but the total crude consumption of all these plants did not equal that of Standard's Bayonne facility. Furthermore about one-fifth of these independent refineries received their crude through Standard pipelines and were thus incapable of truly effective competition. In all, the Bureau of Corporations estimated that Standard refineries processed about 83 percent of the nation's crude in 1906.[6]

Standard was no less dominant in marketing. The oil trust divided the United States into eleven separate marketing territories. In some districts the marketing agent was a department of a Standard firm also engaged in producing, transporting, and refining oil. Standard Oil of California operated in this fashion on the Pacific Coast. In other districts, such as Kentucky Standard's territory in the southeastern states, a separate company did the marketing and purchased refined products from one of the several Standard refineries. Each of the eleven marketing firms operated with the distinct understanding that there would be no competition with the Standard companies in adjoining districts. In 1906 Standard marketing subsidiaries sold over 80 percent of all refined products in the United States and controlled over 85 percent of the export business. After the turn of the century, Standard's position in the marketing of lubricants, fuel oil, and gasoline declined somewhat, but the trust's almost total control of the illuminating oil market continued unimpaired.[7]

Marketing practices were probably the most widely publicized of the oil trust's misdeeds. A seemingly endless succession of journalists and federal investigators detailed these abuses. In 1906 federal investigators documented the following points concerning retail prices: Standard Oil's prices for kerosene varied greatly in different parts of the country; prices were lower where competition existed and unreasonably high where it did not; Standard marketers frequently cut prices below cost in order to drive out competitors and then immediately raised prices after the competition was destroyed. Standard marketers also paid rebates to favored customers as part of the system of price cutting. They obtained information about their competitors' sales pattern through an elaborate espionage system, which was financed by paying bribes to railway employees. They then used this information to slash prices to their competitors' customers and induce those customers to cancel orders from the independents. The oil combination also frequently employed bogus independents—such as the Republic Oil Company —to create the impression of competition where none existed.[8]

Standard's preponderance in the market allowed it to exercise considerable control over the price of refined products, and the combination used that power to raise prices. Kerosene, naptha, and paraffin wax were three of the principal refined products at the turn of the century. From the period 1895-1898 to 1903-1906, the average price of these three commodities increased 46 percent. During that same period general commodity prices in the United States rose only 26.6 percent. Furthermore, rapidly expanding production and Standard pressure stabilized the price of crude oil, while manufacturing costs did not significantly increase. At the same time the combination pursued a strikingly different pricing policy abroad. While American prices for illuminating oils sharply increased between 1895 and 1905, foreign prices were declining. The reason was not an abundance of oil abroad—the amount of oil produced overseas was declining—but a price war between the American trusts and its European competitors. Standard was charging high monopoly prices at home and then using those profits to subsidize price cutting abroad. American consumers, in effect, were financing Standard's quest for new overseas markets.[9]

High retail prices in the United States coupled with stable crude oil prices resulted in increased profit margins. The margin on illuminants for domestic sales was $.053 a gallon from September 1897 through the end of 1899; it climbed to $.060 from 1900 to 1902; and it reached $.066 from the beginning of 1903 through June 1905. The average profit margin of Standard refineries on a gallon of oil processed for all purposes increased from $.009 in 1893 to $.0305 in 1905. The broad scope of the trust's business and these ever-widening profit margins resulted in enormous profits. From 1882 to 1896 the combination averaged a healthy 19 percent annual earnings on capital. After 1897, profits skyrocketed, reaching an annual return of 83 percent in 1903 and averaging 68 percent from 1903 to 1905. Total profits from 1882 to 1906 amounted to something between $790 million to $850 million on an initial investment of $70 million. Dividends for that same period exceeded $500 million, one quarter of which went to John D. Rockefeller.[10]

Prices, profits, and publicity made Standard Oil a prime target for the United States Justice Department. William H. H. Miller,

President Harrison's attorney general, began receiving inquiries about action against the oil combination within a year after the passage of the Sherman Act. In June 1891, G. A. Copeland, editor of the Boston *Daily Advertiser*, wrote to Miller demanding information about federal antitrust enforcement. He inquired whether the department had attempted to gather evidence against "the well-known Standard Oil Trust, the reorganized sugar trust of New Jersey, the cotton oil trust, or any of the better known combinations in restraint of trade." The attorney general cautiously replied that federal district attorneys soon would be "instructed to investigate and prosecute, if they can find any violations of the law."[11]

Attorney General Miller flirted with the idea of action against Standard Oil in August 1891. He ordered a United States district attorney in Ohio to ascertain the accuracy of the "popular belief that its [Standard Oil's] operations are in gross violation of the purposes of the [antitrust] law."[12] That same month, a United States attorney in Missouri requested that Miller supply him with information on Standard Oil gathered by the House Committee on Manufactures during its 1888 investigation. Miller promptly sent the requested material and wrote, "I hope no effort will be spared to prosecute under this law where the facts warrant it."[13] Neither investigation, however, led to litigation.

By 1892 Attorney General Miller had decided against prosecuting Standard Oil. S. R. Kepler's unsuccessful attempt to secure a federal investigation of the oil trust demonstrates the depth of the department's apathy about the competitive situation in the petroleum industry. Kepler had started an oil retailing business in Ashville, North Carolina, in 1887, but the oil trust refused to supply him with petroleum products and resorted to various anticompetitive practices to drive him from business. In June 1890, Kepler bought space in the Raleigh *State Chronicle* and publicly called on the governor of North Carolina to take action against Standard Oil under provisions of the state's recently enacted antitrust law. North Carolina officials were less than eager to help, so Kepler turned to Washington. On May 13, 1892, he wrote to Attorney General Miller, inquiring "how to obtain a conviction by the U.S. Govt. of the Standard Oil Trust under the Antitrust Law." Miller replied that Kepler should refer the matter to Charles Price, the United States attorney for western North Carolina.[14]

On July 16, 1892, Kepler explained to Attorney General Miller the problem with this course of action:

I beg to say that I have already applied to Mr. Price both in person and by newspaper publication and was in no measure encouraged to expect any help from him in the matter of prosecution of Standard Oil Co. Mr. Price is besides U.S. Atty. also Atty. for the Richmond & Danville Rail Road Co., and is in this capacity a beneficiary and in a certain sense an employee of the Standard Oil Co.—As the Standard Oil people are large stockholders in the Richmond and Danville Co. . . .Can you not suggest something to me outside or apart from Mr. Price as a prosecution.[15]

Miller filed this request, and Kepler's next one dated August 8, 1892, without response.[16]

On August 17, an exasperated Kepler again demanded action, this time threatening, "Not receiving reply within five days the only means left to get your attention will be through the public prints." The Justice Department was forced to respond. On August 20, the acting attorney general informed Kepler that "it would not be prudent to institute other proceedings until the law has been thoroughly tested in the suits now pending." The department, however, reminded Kepler that he still had the right to proceed under section 7 of the Sherman Act, which allowed a private citizen whose business or property had been adversely affected by an illegal combination to file suit in a United States circuit court to "recover threefold the damages sustained by him."[17] But a private suit against the oil trust presented great problems. Gathering evidence was both time-consuming and expensive, and Standard Oil would be represented in court by outstanding lawyers, who were capable of dragging out the legal process for years. Understandably, Kepler did not pursue the suggestion.

President Cleveland's first attorney general, Richard Olney, displayed a similar lack of enthusiasm for legal action against the oil trust. In June 1894, a group of independent petroleum producers, refiners, and exporters from New York, Pennsylvania, Ohio, and West Virginia held a convention at Warren, Pennsylvania. They exchanged evidence of Standard Oil's anticompetitive practices and unanimously adopted a resolution calling for state and federal antitrust action to halt "the growing arrogance of this monster monopoly."[18] The oil men sent a copy of the resolution to the Justice Department with a request for action. Attorney General Olney, following what was to become a departmental tradition, filed the request without reply. In 1893 Olney became secretary of state, and Cleveland appointed Judson Harmon attorney general. Undisturbed by further complaints against Standard Oil, Harman took no steps against the combination.

The department preserved its record of inactivity during the McKinley administration. Attorney General Joseph McKenna, who took office in March 1897 and served less than a year, received no complaints against Standard Oil. But John W. Griggs, who assumed the top post at Justice in early 1898, presided over the federal government's most determined effort to ignore a citizen seeking action against Standard Oil. Griggs refused to act on the repeated requests of George Rice, the oil trust's most durable adversary. Rice began his campaign to secure federal antitrust activity against Standard Oil with a letter to Griggs on November 5, 1899. He appealed "not only for individual relief, but also in behalf of thousands of my confreres in the oil producing and refining business, who are sorely oppressed by the most gigantic and unlawful combination the world has known, embodied in the name, Standard Oil Trust." Griggs replied that Rice had not adequately demonstrated the interstate character of the oil combination.[19]

Rice did not discourage easily. On January 10, 1899, he sent Griggs an elaborate sixteen-page statement demonstrating in detail that Standard Oil was indeed an interstate operation. Rice noted that the oil trust's vast pipeline system transported great quantities of oil across state lines. Rice also pointed to Standard Oil's close operating relations with various interstate railroads. The Justice Department filed Rice's document without reply. On February 16, 1899, Rice sent a twenty-page letter on the same subject. Finally, on March 14, 1899, Griggs acknowledged receipt of both letters and claimed they were receiving "proper consideration." But the attorney general stubbornly added: "It does not appear from your communication that either your business or that of the alleged combination against which you complain is of an interstate character such as to give the Courts of the United States jurisdiction under the Statute mentioned.''[20]

A disgusted Rice responded on April 17, 1899: "There is nothing intricate in this case, it is about as plain and transparent as daylight . . .and with the abundant proof in your hands farther [sic] delay in the consideration of so plain a case is inexcusable and entirely unnecessary."[21] The Justice Department did not reply, nor did it respond to Rice's complaints of April 24, May 1, and May 15.[22] On May 5, Rice wrote to President McKinley, charging that the Justice Department had refused to act on his complaints for six months and that his business was being destroyed: "I appeal to you, and call your attention thereto, that farther [sic] delay or leniency should not be allowed, that summary action long ere this should have been taken by the Government to rid this country of the baneful effects of this trust."[23] McKinley referred Rice's letter to the Justice Department, where it, like the others, was filed without reply.

Next Rice had his lawyer, A. G. Stafford, attempt to get a response from the Justice Department. On May 31, 1899, Stafford wrote to Griggs, demanding action on Rice's numerous complaints. When the department declined to answer his letter, Stafford wrote again on July 19, insisting that he receive an explanation.[24] Attorney General Griggs then offered the following reason for the department's lack of activity:

The matter referred to has been heretofore in the charge of Solicitor-General Richards, who has had general charge of violations of the Sherman act for this Department. Mr. Richards is at the present time out of the country, and I have not been able either to refer the matter to him or to confer with him relative to the subject, which, as you must be aware, is one requiring careful investigation and consideration.[25]

Stafford made a final request for action on July 26, 1899, and received no reply.[26]

At the hearings before the Industrial Commission in late 1899, independent oil men summed up their experience with the Justice Department. Lewis Emery, who had participated in the Warren,

Pennsylvania, conference of 1894, complained of the Justice Department's failure to enforce the Sherman Act: "Complaint of our difficulties was made during the Cleveland administration to the Attorney-General of the United States, and no answer was made by the receipt of the letter."[27] George Rice was also on hand to relate his experiences: "I have had correspondence with the Attorney-General of the United States myself in regard to proceedings against the Standard Oil Trust, but he doesn't act." He contemptuously added that the attorney general "made excuses from one thing to another; that he hadn't time to look into it, and the Assistant Attorney had gone to Europe and hadn't got back."[28]

Apart from the indifference of many top Justice Department officials, there were several reasons for their refusal to act against Standard Oil. In 1890 the department's Washington office had a modest staff of eighteen lawyers and about sixty clerical workers, and staff size did not substantially increase over a decade. Departmental offices were located on the upper floor of the dilapidated Freedman's Bank Building, located on Pennsylvania Avenue across from the Treasury Department. Working conditions were so poor that Congress finally was forced to provide funds for relocating the department in 1899. The attorney general and his staff then moved to the Baltic Hotel on K Street and Vermont, and the Freedman's Bank Building was demolished. Throughout the final decades of the nineteenth century, the small, inadequately housed staff faced a rapidly expanding workload: the annual number of cases before the Supreme Court tripled during the 1880s, and the number of suits in the court of claims increased twentyfold.[29]

By passing the Sherman Act, Congress assigned the already overburdened Justice Department the formidable task of policing the national economy without providing the special funds or the additional personnel necessary to carry out the task. In 1896, the House of Representatives, responding to public displeasure over the Justice Department's poor record in antitrust enforcement, asked Attorney General Judson Harmon to account for departmental inactivity. Harmon justifiably pointed to the lack of resources: "If the Department of Justice is expected to conduct investigations of alleged violations . . . it must be provided with a liberal appropriation and a force properly selected and organized. The present appropriation for the detection of crimes and offenses is very small, and the time of examiners is fully occupied by the present important duties assigned to them."[30]

The limited resources of the Washington office of the Justice Department and the far-flung operations of the larger trusts made cooperation of the various United States district attorneys essential to effective antitrust enforcement, but the attorney general did not effectively control his subordinates in the field. The government's method of paying the district attorneys caused much of the difficulty. These officials earned a fixed salary of two hundred dollars per year, but the bulk of their income—about four thousand dollars in 1891—came from fees paid by the government based on the number of cases conducted. The district attorneys thus frequently avoided antitrust cases because they were so time-consuming. The fee system ended in 1896, but it had contributed to the weakness of federal antitrust enforcement during the first years of the Sherman Act.[31]

In the face of these assorted difficulties, the Justice Department actively investigated only a small number of the complaints it received. To make matters worse, numerous problems plagued the few antitrust probes that were launched. Government attorneys had to analyze the operations of complex industrial organizations, a task they were ill trained to perform. Federal investigators faced the same difficulties in extracting information from corporations that were so apparent in the state litigation against Standard Oil. Corporate officials could refuse to provide material, and the Justice Department did not have the power of subpoena during the initial investigation. Businessmen also could open their files and swamp investigators in a sea of irrelevant material. Whatever the tactics, the result was the same: investigations spanning years and consuming the department's money and manpower.[32]

The Sherman Act authorized the Justice Department to bring criminal charges against corporate directors or agents and to file suits in equity against combinations in restraint of trade. Both remedies had advantages, but they also had serious drawbacks. A criminal suit offered the prospect of dramatic punishment for past violations that might deter others. A successful criminal prosecution in an important antitrust case, however, required the government to convince both a grand jury and a trial jury that a successful businessman was guilty of violations in matters often obscured by complex economics. A suit in equity offered an easier path to conviction, and a court decree resulting from a successful prosecution could dismantle the offending combination and restore competition. Yet the decree could not reach profits illegally earned in the past. It could prohibit certain practices, but businessmen and their lawyers remained free to devise methods of working around those prohibitions. Moreover, the courts lacked power to guarantee compliance with their directives. If the Justice Department discovered a subsequent violation of a court decree, its only resource was to begin the same defect-ridden process again.[33]

The important 1895 United States Supreme Court decision in *United States* v. *E. C. Knight Company* also discouraged aggressive antitrust enforcement. In 1892 the American Sugar Refining Company, already manufacturing over 60 percent of all refined sugar in the United States, dramatically increased its hold on the industry by purchasing the E. C. Knight Company and three other sugar-refining firms. American Sugar's four new acquisitions together produced over 30 percent of the nation's refined sugar, and the government, charging that these purchases constituted a combination in restraint of trade, brought suit to compel cancellation of the sales contracts.[34]

The Supreme Court ruled against that suit. Chief Justice Melville W. Fuller, speaking for the majority, differentiated sharply between manufacturing and commerce. The Sherman Act applied only to combinations restraining interstate commerce. The law did not cover manufacturing combinations like the sugar trust because manufacturing was local in character and only indirectly affected interstate commerce. Therefore, even though the defendant combination controlled the manufacture of over 50 percent of all refined sugar in the United States, it was not illegal. This decision placed a great number of the nation's largest corporations beyond the scope of the Sherman Act. It was a devastating blow to an effective federal antitrust policy.

Toward the end of the decade, however, the Supreme Court handed down a series of decisions that blocked further erosion of the federal antitrust law. In the 1897 case of *United States* v. *Trans-*

Missouri Freight Association, the court upheld a government attempt to dissolve a rate-fixing agreement between eighteen previously competing railroads in the Southwest. The defense attorneys argued that the Sherman Act did not apply to railroads because they were already under the jurisdiction of the Interstate Commerce Commission. They further contended that the federal antitrust law allowed reasonable restraint of trade and that the Trans-Missouri agreement was reasonable because it brought stability to the ruinously competitive railroad industry. But Justice Rufus W. Peckham, speaking for a five-man majority, upheld the government on both issues. He declared that the language of the Sherman Act made "every contract, combination . . . or conspiracy, in restraint of trade" void. Therefore, in the absence of compelling evidence to the contrary, railroads must be included within the provisions of the law. Peckham also ruled that the Sherman Act pronounced all contracts or combinations in restraint of trade—reasonable or not—void.[35]

Less than a year later, the Supreme Court again sustained a government effort to apply the Sherman Act to a railroad association in *United States* v. *Joint Traffic Association.* The case involved an association of thirty-one railroads that had combined to fix rates and fares between Chicago and the East Coast. Railroad attorneys contended that certain technical differences between this agreement and the Trans-Missouri combination made the two fundamentally different. Justice Peckham, speaking for a five-man majority, found these distinctions unimportant. He ruled that the "natural and direct effect" of both agreements was to maintain high rates and restrain interstate trade. Railroad attorneys also attempted to pursuade the court to reconsider the Trans-Missouri decision and allow reasonable restraints on trade, but Peckham refused.[36]

In 1899, the Supreme Court handed down still another decision supporting a government prosecution under the Sherman Act. *United States* v. *Addystone Pipe & Steel Company* involved an association of six cast-iron manufacturers, controlling nearly one-third of the nation's production capacity. The association awarded bidding rights on particularly large orders to the member company that offered to pay the largest bonus to the association. The other companies would then protect the winning company by submitting slightly higher bids. Each firm had exclusive marketing areas for orders of moderate size. The association periodically distributed the bonuses it had accumulated to the member companies.[37]

Defense lawyers argued that the association was essentially a manufacturing combination and was thus beyond the scope of Congress's power to regulate interstate commerce under the *Knight* rule. Justice Peckham, for the first unanimous Court in a Sherman Act case, strongly rejected that proposition. He ruled that "contracts for the sale or transportation to other States of specific articles were proper subjects for regulation because they did form part of such [interstate] commerce."[38] Since the cast-iron manufacturers association involved just such a contract, it was void under the Sherman Act. *Addystone* was the first Supreme Court decision since *Knight* involving an industrial combination. Although manufacturing concerns remained beyond the scope of the Sherman Act, the decision allowed the Justice Department to proceed against old-fashioned pools operating across state lines. Federal prosecutors, however, displayed little enthusiasm for exploring the possibilities presented by this decision.

Theodore Roosevelt became president of the United States on September 14, 1901, when President McKinley was assassinated. Unlike his three predecessors, Roosevelt fully appreciated the importance of the trust problem. He had first seriously confronted the issue while serving as governor of New York. In 1899, Roosevelt expressed alarm at "the growth of popular unrest" over the rising power of monopoly. He feared that if the established political leadership did not develop "some consistent policy to advocate then the multitudes will follow the crank who advocates an absurd policy, but who does advocate something."[39] Significantly, Roosevelt's concern with monopoly was defensive. He needed a coherent position on the issue to prevent "crank" reformers from building support for solutions that might threaten the existing power structure. He displayed little real concern about the serious social and economic problems created by industrial combination. Roosevelt admitted that there was "a great deal of misery and injustice" associated with modern industrial conditions, but he attributed those difficulties to "the faults of the individuals themselves, or to the mere operation of nature's laws."[40]

Roosevelt's approach to the trust question involved several central propositions. First, he accepted the widespread, but erroneous, notion that combination was an unavoidable component of industrial development:

Much that is complained about is not really the abuse so much as the inevitable development of our modern industrial life. We have moved far away from the old simple days when each community transacted all its work for itself and relied upon outsiders for but a fraction of the necessities, and for not a very large part of the luxuries, of life.[41]

Nevertheless, he acknowledged that in some instances corporate misconduct was too flagrant to be passively endured by the public: "The chicanery and the dishonest, even though not technically illegal, methods through which some great fortunes have been made, are scandals to our civilization." He pointed to secrecy, overcapitalization, and exorbitant prices as leading examples of corporate malfeasance.[42]

Believing that combination was the irresistible wave of the future and that industrial abuses were the result of moral inadequacies, Roosevelt concluded that drastic measures were not necessary to solve the problem. He thought that legislation to end corporate secrecy might work: "The first essential is knowledge of the facts, publicity."[43] If stockholders and consumers knew about corporate misbehavior, businessmen would have a powerful incentive to operate ethically, and lawmakers would possess the information necessary for further constructive legislation. Roosevelt also thought that continuous administrative regulation of corporations was essential. "What is needed," he declared in 1905, "is not sweeping prohibition of every arrangement, good or bad, which may tend to restrict competition, but such adequate supervision and regulation as will prevent any restriction of competition from being to the detriment of the public."[44]

Roosevelt had little faith in antitrust laws as a basic tool for dealing with big business. He believed they had proven "absolutely ineffective" because they attempted to ban the natural course of industrial development. They provided only sporadic action against the trusts, while modern industry required continuous supervision. As Roosevelt put it, "Much of the legislation not only proposed but enacted against trusts is not one whit more intelligent than the medieval bull against the comet, and has not been one particle more effective."[45] Still, the antitrust suit could serve as a means of publicizing information about corporate abuse while demonstrating the government's concern about the monopoly problem. The Sherman Act, a singularly defective instrument of public policy, was thus pressed into service in President Roosevelt's campaign to neutralize the trust issue without disturbing the existing political and economic power structure.

Philander C. Knox, Roosevelt's first attorney general, drafted and secured several bills to prepare federal antitrust machinery for its new role. In 1903, Congress approved the Expediting Act, which allowed the attorney general to certify to circuit court judges that a specific equity suit under the Sherman Act was of "general public importance." The case would then be given precedence over all others and could be appealed only to the United States Supreme Court. That same year Congress passed the Antitrust Appropriation Act, granting the Justice Department $500,000 for fiscal 1904 to finance enforcement, and the Deficiency Act, authorizing the attorney general to appoint two high-level officials to deal exclusively with antitrust cases. The Deficiency Act marked the beginning of the Antitrust Division of the Justice Department.[46]

Knox began only five antitrust suits during his four years as attorney general.[47] His suit against the Northern Securities Company resulted in the most notable government antitrust victory up to that time. Owners of the Northern Pacific Railway and the Great Northern Railway, two competing lines in the Northwest, incorporated the Northern Securities Company in New Jersey in 1901. The new corporation, capitalized at $400 million, served as a holding company. It transferred its stock to Northern Pacific and Great Northern, and in return it received a controlling interest in both firms. Because the transaction effectively ended competition between the two northwestern railroads, Roosevelt and Knox decided that antitrust action was required. The government filed an equity suit in 1902 to dissolve the Northern Securities Company as a combination in restraint of trade. The case reached the Supreme Court in 1903.

The court, in a five to four decision, ruled that the formation of the Northern Securities Company had violated the Sherman Act. Justice John M. Harlan, who wrote the majority opinion, rejected the defense contention that the holding company was the result of a mere stock transaction and therefore did not involve interstate commerce. Harlan ruled that a stock transfer was illegal if it "directly or indirectly" caused a restraint of commerce. In the present case, the two railroads "have become, practically, one powerful consolidated corporation, by the name of a holding company, the principal, if not the sole, object for the formation of which was to carry out the purpose of . . . combination under which competition between the constituent companies would cease."[48] This decision was particularly significant because it broadened the definition of commerce and thus modified the extremely narrow interpretation of that term advanced in *Knight*.

Roosevelt was delighted with this highly publicized victory. When Knox left the Justice Department in June 1904, the president lavishly praised his departing attorney general's antitrust record: "Under you it had been literally true that the mightiest and the humblest in the land have alike had it brought home to them, that each was sure of the law's protection while he did right, and that neither could defy the law if he did wrong."[49] But independent oil dealers would have taken exception to the president's glowing praise.

George Rice had begun his campaign to convince Attorney General Knox to take action against Standard Oil while McKinley was still president. On June 26, 1901, Rice wrote to Knox: "I take it for granted that you are aware, or that you have been informed ere this, or since your appointment, that more than two and a half years have elapsed (Nov. 5th, 1898) since I entered general complaint against the Standard Oil Trust."[50] Knox did not reply, and Rice wrote again on August 9 demanding an answer. Six days later

Knox responded, claiming that Rice's original letter had taken an unusually long time to reach him and that the matter was under consideration.[51]

Although he suspected that Knox would adopt the same policy as his predecessors had, Rice had not totally given up hope. On August 16, 1901, he wrote: "I shall endeavor to address you in respectful language, and most sincerely hope that in my future relations with yourself that I may receive from your department respectful, prompt and due consideration of the all important matter presented."[52] Five months later, in January 1902, Rice supported his plea for action with an elaborate seven-page document recounting his repeated attempts to prompt federal action and detailing Standard's anticompetitive practices.[53]

On January 9, 1902, Knox sent his only substantive response to Rice's repeated pleas. Knox took the same position Griggs had embraced. The Justice Department had not acted against Standard Oil because Rice's complaints "did not present sufficient grounds upon which to base the proceedings requested." Specifically, Standard Oil did not appear to be engaged in interstate commerce. Knox could only suggest that Rice file suit under section 7 of the Sherman Act: "If your complaint is true, and you can prove it, you ought to have no difficulty in securing competent counsel in Ohio and ample redress in the Federal courts there."[54]

Rice made his final assault on the Justice Department in the summer of 1902. He began on July 10, with a thirty-four-page letter, complete with a collection of newspaper clippings, demonstrating the interstate character of the oil trust. Rice added, "In view of your present procedure against Northern Securities Company and the Beef Trust it becomes pertinent for me to inquire why it is even at this late day no proceedings have been instituted by the Department of Justice against the well-known public violator of our laws the Standard Oil Trust?"[55] By this time, Knox had had enough of George Rice. The Attorney General filed the letter, and two more that Rice wrote on August 16 and 17, without response.[56]

About the same time, Rice decided to test Theodore Roosevelt's interest in the oil situation. On August 1, 1902, he informed the president of the Justice Department's repeated refusal to take action on his complaints. Rice concluded, "It is quite obvious that your Attorney General does not intend to take cognizance of these the most important and the most extraordinary of complaints that were ever presented to the Department of Justice, dating back as far as Nov. 5th, 1898.''[57] Although Rice pleaded for presidential intervention to force an investigation of the oil trust, Roosevelt merely referred the letter to the Justice Department, where it was filed without reply. During the next three months Rice sent the president seven more letters in which he quoted the language of the Sherman Act and Roosevelt's own antimonopoly pronouncements. The president refused to acknowledge receipt of these messages. Like Rice's first letter, they were sent to the Justice Department and buried in the files.[58]

Finally, on December 30, 1902, Rice made his final appeal to the White House:

I wrote to you on August 1st, the 8th, the 15th, the 22nd, and the 29th; also on September 5th and the 12th; also October 28th; and this is my ninth appeal to you, without recognition or response to either one of my eight previous communications, in relation to the most extraordinary of complaints, irrefutable charges, not one of them questioned or denied, which I have made to the Department of Justice. . . . Your Attorney General refuses to consider or take action upon these complaints, therefore, I have appealed to you as the court of last resort.[59]

But the president still revealed no interest in an oil investigation. Again he referred the letter to the Justice Department without comment.

George Rice was not the only one filing complaints against Standard Oil during Roosevelt's first term. A member of the Indiana Oil Men's League claimed that Standard Oil was a monopoly and should be destroyed, just as the Northern Securities Company had been broken. A Minneapolis oil dealer informed the Justice Department that Standard was cutting prices to destroy competition. An Indiana oil dealer wanted a Justice Department investigation because the oil trust had cut off his oil supply. A West Virginia man wrote to President Roosevelt, offering to give testimony against Standard Oil. And a San Francisco man asked Roosevelt to launch an investigation of Standard Oil of California's relations with the railroads.[60] But the administration tenaciously adhered to its policy of inaction.

On June 27, 1904, after unsuccessfully trying to get federal action for six years, George Rice finally filed a private suit under section 7 of the Sherman Act in the federal circuit court in New Jersey. He contended that the Standard Oil trust agreement of 1882 was an illegal contract and formed an illegal combination in restraint of trade and claimed that Standard Oil had destroyed his business, valued at $750,000. The oil trust had induced the railroads to charge him discriminatory and exorbitant rates to ship his products and then demanded that the railroads pay it the excess rates extracted from Rice. Standard also compelled the railroads to delay Rice's shipments and to refuse to assign him railroad cars when he needed them. Finally, the combination sold oil to Rice's customers at less than cost in an effort to drive him from business.[61]

Standard attorneys argued that Rice's declaration was defective, and on January 6, 1905, circuit court Judge William B. Lanning upheld that position. He ruled that Rice's declaration "must aver not only facts showing such a contract or combination or conspiracy as is declared by the act to be unlawful, but facts showing that by reason of such unlawful thing he had been injured in his business or property."[62] In short, Rice had not made his charges specific enough. He had not demonstrated what railroad rates were exorbitant, what shipments had been delayed, or where and when Standard Oil had sold oil below cost. Judge Lanning concluded that "the averments in the declaration are too vague to give the defendant the information to which it is entitled before being required to plead."[63]

Rice's suit failed because he lacked the financial resources to conduct the wide-ranging investigation necessary to substantiate his charges against Standard Oil. He knew that his charges would be difficult to document, which is why he had tried to convince the federal government to act. When the Roosevelt administration refused to move, Rice was forced to fight the oil trust alone. He died three months after his defeat in court. The New York *World* observed, "George Rice, who gave the declining years of his life to a relentless fight against the Standard Oil Company in Ohio, which he hoped to dissolve, lies dead in Asbury Park. Worn in mind and body, his end was hastened by an adverse decision in a suit for $3,000,000 against the trust."[64]

Roosevelt was not interested in citizens' requests to enforce the Sherman Act, but he understood the political significance of Standard's poor public image. In January 1903 he demonstrated how skillfully he could manipulate the symbol of the oil trust while seeking trust legislation giving statutory form to his passion for publicity, continuous administrative supervision, and executive discretion. In particular he backed a bill creating a department of commerce and labor that included a bureau of corporations. Through an amendment drafted by Attorney General Knox, this bureau received the power to compel testimony and force corporations to produce documents. Furthermore publication of bureau reports was to be entirely at the president's discretion. This last provision gratified Roosevelt, but many congressman viewed it as an unwarranted extension of executive power.[65]

During the first week of February, the bill faced strong opposition in Congress. On the sixth, the oil barons unwittingly supplied the president with more than enough ammunition to crush his opponents. John D. Archbold sent a telegram to Senator Matthew Quay of Pennsylvania:

We are unalterably opposed to all proposed so-called Trust bills except the Elkins Bill already passed by the Senate, preventing railroad discrimination; everything else is utterly futile and will result only in vexatious interference with the industrial interests of the country. The Nelson bill [creating the Bureau of Corporations] as all others of like character, will be an engine for vexatious attacks against a few large corporations.[66]

That same day, John D. Rockefeller, Jr., sent a message to several other senators:

Our people are opposed to all proposed trust legislation except the Elkins Anti-Discrimination Bill. Mr. Archbold, with our counsel, goes to Washington this afternoon. Am very anxious they should see you at once and shall much appreciate any assistance you can render them.[67]

Upon learning of these communications, Roosevelt immediately set up a meeting with reporters. He informed the assembled newspapermen that John D. Rockefeller himself had sent telegrams to six senators, declaring in substance: "We are opposed to any antitrust legislation. Our counsel, Mr. ———, will see you. It must be stopped."[68] Although the reporters were unable to locate any senators who would acknowledge having received a message from Rockefeller, Sr., Roosevelt's story produced the desired effect. The papers printed the president's account on February 8; the House passed the Bureau of Corporations bill by a vote of 252 to 10 on February 10; and the Senate enacted the measure without opposition the following day. Roosevelt signed the bill into law on February 14.[69]

The president was pleased with his work. "I got the bill through," he exclaimed, "by publishing those telegrams and concentrating public attention on the bill."[70] To achieve that end, he falsely attributed the telegrams to Rockefeller, Sr., and sharpened the language of the messages. But more important, Roosevelt demonstrated that he was both keenly aware of the potent political symbolism surrounding John D. Rockefeller and the oil trust and capable of using that symbolism to his political advantage.

Roosevelt's next opportunity to exploit the Standard image occurred during the 1904 presidential election and involved the work of the Bureau of Corporations. The bureau accomplished very little during its first year, and Roosevelt did not want to press antimonopoly activity during the election because he relied heavily on corporate contributions to finance his campaign. Fully 72 percent of Roosevelt's $2,195,000 campaign chest came from large corporations. He received $150,000 from J. P. Morgan, $50,000 from C. S. Mellon, and $50,000 from E. H. Harriman. Standard Oil also contributed generously. Early in September 1904, Cornelius N. Bliss, the Republican National Committee treasurer, solicited funds from Archbold and Rogers. After being assured that the contribution was "thoroughly approved by the powers that be," the oil trust gave $100,000.[71]

On October 1, 1904, the New York *World* accused the president of rampant hypocrisy on the trust issue. Noting that the Bureau of Corporations had produced nothing in its first year and a half of operation, the paper observed that Roosevelt's campaign workers were busily amassing a fortune in corporate contributions. The *World* then advanced the plausible but unsupported thesis that the contributions were payment for the inactivity of the bureau.[72]

Roosevelt, who recognized the political imperative of keeping a safe distance from the trusts in public, moved quickly to protect himself. In need of a particularly grand gesture, he ordered Republican National Committee chairman George B. Cortelyou to return Standard Oil's contribution on October 26. William Howard Taft later recalled the denouement of this episode:

[Attorney General] Knox said he came into the office of Roosevelt one day in October, 1904, and heard him dictating a letter directing the return of $100,000 to the Standard Oil Company. He said to him, "Why, Mr. President, the money has been spent. They cannot pay it back—they haven't got it," "Well," said the President, "the letter will look well on the record, anyhow," and so he let it go.[73]

The GOP kept the $100,000, and Treasurer Bliss requested an additional $150,000 from Standard Oil; the oil barons refused.[74]

Shortly after Roosevelt's reelection in 1904, the Bureau of Corporations began an investigation into the oil industry—a probe that eventually provided the president with his third opportunity to capitalize on the unpopularity of Standard Oil. This inquiry grew out of the troubled conditions in the Kansas oil fields and focused on the combination's abuse of petroleum transportation systems. Standard officials, believing that they had little to fear from the toothless bureau, cooperated. But James R. Garfield, commissioner of corporations, was determined to make a searching investigation because his 1905 report on the beef industry had been violently denounced by the press as a whitewash. Even worse, the beef report had not impressed the president.[75]

Garfield did not intend to make the same mistake twice. On May 2, 1906, the bureau issued an exposé of the oil trust's machinations with the railroads under a deceptively bland title, *Report on the Transportation of Petroleum*. Federal investigators demonstrated that Standard routinely received illegal secret rates on oil shipments and that these covert arrangements saved the combination $750,000 in 1904 alone. The intimacy between the trust and the railroads was well developed because the special favors were "so secret, so ingeniously applied to new conditions, and so large in amount." The shippers even granted the oil combination preferential treatment in openly published rates.[76]

Roosevelt knew how to use the report. The debate on the Hepburn Act, granting the Interstate Commerce Commission increased powers over railroad changes, had reached a critical point in the Senate. On May 4, 1906, the president summarized the bureau's conclusions in a special message to Congress. He emphasized that Standard Oil derived tremendous competitive advantage from its system of discriminatory transportation rates. He observed that the bureau's report was "of capital importance in view of the effort now being made to secure . . . enlargement of the powers of the Interstate Commerce Commission." And he implied that the Hepburn Act would go a long way toward ending those abuses. Roosevelt's timing was superb; Congress responded to his message by promptly passing the bill.[77]

By focusing public attention on the report, Roosevelt increased the already intense public pressure for a federal antitrust suit against Standard Oil. Newspapers showered abuse on the combination and called for legal action. The New York *Sun* declared: "Every power of the Federal Government, every foot-pound of energy in the Administration, should be directed fearlessly through the Department of Justice for the prosecution of the beneficiaries, both givers and takers, of the enormous system of secret rates described in Commissioner Garfield's report." The Boston *Daily Advertiser* demanded that the government base an antitrust suit on Garfield's evidence. And the Rochester *Herald* asserted that Standard Oil should be considered guilty until proven innocent. Meanwhile, complaints from throughout the country continued to pile up in the Justice Department.[78]

Roosevelt, who had studiously ignored a steady stream of citizens' complaints against the oil combination during the first five years of his administration, now judged the political situation right for antitrust action against Standard. His subsequent handling of the case would demonstrate repeatedly that public opinion, not the defect-ridden antitrust process, was his real concern. On June 22,

1906, William H. Moody, who had replaced Philander Knox as attorney general in 1904, announced that the Justice Department was beginning a preliminary investigation in an antitrust suit against Standard Oil of New Jersey. Understandably, Moody rejected William H. Taft's suggestion that the government acknowledge the mountain of complaints already received against Standard Oil.[79]

The attorney general asked Frank B. Kellogg, who had recently worked on a government antitrust probe of the paper trust, to take charge of the investigation. Kellogg had served as counsel for the United States Steel Corporation, and he had profound respect for property rights. Before the National Civic Federation's 1907 conference on trusts, he declared that "the right of property and its control cannot be too sacredly guarded." But Kellogg, like Roosevelt, feared that unrestrained corporate abuses might radicalize the public and endanger the existing order. He asked those attending the conference:

Do you wish to drive this people into Socialism, where they will compel the government to take possession of and manage the commerce, the manufactures, of all the industries, the cultivation of the soil for the benefit of all? If you do not wish to do this, then put a stop to the power of unlimited combination, and do it by orderly means.[80]

Antitrust litigation under the Sherman Act was the "orderly means" Kellogg had in mind. Attorney General Moody also appointed Charles B. Morrison, then federal district attorney in the Northern District of Illinois, to the case. Morrison had gained experience in antitrust litigation by working on the federal government's suit against the beef trust.[81]

Kellogg and Morrison, who proved to be conscientious investigators, had little difficulty massing evidence against Standard Oil. Garfield consulted regularly with the special prosecutors and provided access to the Bureau of Corporation's wealth of material on the oil trust. The Interstate Commerce Commission contributed an eight-volume study on Standard's relations with the railroads, and George Rice's daughter even offered to provide evidence against her late father's adversary. By early September the special prosecutors completed an initial report that judged the Standard organization a combination in restraint of interstate commerce. Focusing on organizational structure, Kellogg and Morrison concluded that the Jersey Standard holding company violated the Sherman Act under the precedent established in the Northern Securities case. Roosevelt instructed Moody to proceed with the case.[82]

On November 15, 1906, the prosecutors filed a suit in equity against Standard Oil of New Jersey, John D. Rockefeller, John D. Archbold, Henry M. Flagler, four other directors, and various subsidiary corporations of the combination. The bill charged that the combination was a monopoly and engaged in a conspiracy in restraint of trade within the prohibitions of the federal antitrust law. The government requested that the Jersey Standard directors be enjoined from exercising control over subsidiary corporations and that the subsidiaries be prohibited from paying dividends to the holding company. The prosecutors chose their point of attack carefully. They presented their case to the same federal circuit court at St. Louis that had handed down the original ruling against the Northern Securities holding company in 1903.[83]

The oil barons assembled a distinguished defense team to fend off the antitrust challenge. John G. Milburn, one of the leading lawyers in New York, served as chief counsel. John G. Johnson, who twice refused nomination to the United States Supreme Court, and David T. Watson, who had worked for the government in the Northern Securities case, also appeared on behalf of the trust.[84] The defense first attempted to have the case transferred out of the St. Louis circuit court. Shortly after the prosecutors submitted their bill in equity, the court issued subpoenas to Standard officials throughout the nation. On December 26, 1906, the Standard attorneys filed a petition to set aside the service of process outside the district of the St. Louis circuit court on grounds that the court lacked nationwide jurisdiction. The defense argued that the government had to file suit in the district where the principal defendant, Jersey Standard, resided. Even if this jurisdictional challenge failed, it would slow down the progress of the case, and delay was an important oil trust tactic in combating antitrust litigation.[85]

At this point, Charles J. Bonaparte succeeded William Moody as attorney general and assumed responsibility for the case. In an attempt to minimize the delay caused by the jurisdictional dispute,

Bonaparte declared the Standard case to be of general public importance and placed it under the provisions of the Expediting Act of 1903. The St. Louis circuit court then promptly scheduled a hearing, and on March 7, 1907, the court ruled against the Standard attorneys. Judge Walter H. Sanborn quoted section 5 of the Sherman Act: when "the ends of justice require that other parties should be brought before the court, the court may cause them to be summoned, whether they reside in the district in which the court is held or not." He then ruled that the Waters-Pierce Oil Company, a resident of Missouri, was a key member of the alleged nationwide conspiracy to monopolize the oil industry and that the "ends of justice" required the court to hear testimony from individuals throughout the country.[86]

After turning back the jurisdictional challenge, Kellogg and Morrison energetically plunged into a massive investigation of Standard's position in the petroleum industry, unhampered by the lack of resources that frequently plagued other antitrust inquiries. Beginning on September 17, 1907, the government prosecutors took testimony in New York, Washington, Chicago, Cleveland, and St. Louis over an eighteen-month period. The prosecution and the defense called a total of 444 witnesses to the stand and introduced 1,374 exhibits. The final record extended to 14,495 printed pages in twenty-three volumes. The investigation was by far the largest antitrust probe to that time. The Bureau of Corporations, which had just concluded its report on the oil industry, provided the prosecution with substantial evidence on the economic aspects of the case. The prosecutors also employed evidence from various state suits against the oil trust, particularly the Ohio cases of the 1890s and the Missouri case of 1906.[87]

While the investigators plowed through a thicket of legal and economic complexities, Roosevelt repeatedly demonstrated his contempt for the antitrust process. In the summer of 1907, both Kellogg and Bonaparte were frustrated by the slow pace of the antitrust suit. They were particularly concerned because the investigation would require extensive testimony from witnesses throughout the country. In June 1907 Kellogg attempted to deal with this problem by having four examiners appointed to take testimony in the case, but the court appointed only one. In August, Bonaparte informed the president of the situation: "If the testimony is taken before a single examiner, he must sit successively in almost every part of the country, since no witness can be summoned by him for more than one hundred miles, the case for the Government will certainly not be completed within the next year, and it will be altogether impracticable to get to a hearing before the close of your administration."[88]

A criminal suit offered a possible way around this problem. The prosecutors had considered filing such a suit as early as July 1906 because of the strong evidence against the Standard leaders.[89] Now this approach presented the additional attraction of speeding up the civil case. Bonaparte explained the prosecutors' plan to Roosevelt:

Various expedients have been considered by which time might be saved, but the only thoroughgoing and certainly effective one would be to institute a criminal proceeding against the corporation, and some of the more prominent individuals who are defendants to the present proceedings, summon the witnesses before the grand jury in such a proceeding, which can be done without regard to place of residence, and then immediately re-summon them, while in the jurisdiction selected, to testify before an examiner in the equity suit.[90]

The oil barons were genuinely concerned about potential criminal action against them, and Standard attorneys proposed an out-of-court settlement to Kellogg and Morrison. But Attorney General Bonaparte advised against compromise. Roosevelt agreed: "If we have a criminal case against these men, I should be very reluctant to surrender it."[91]

The threat of criminal indictment was so serious that Standard officials could not drop the matter. In late September 1907, Fred Goff, a Standard attorney, discussed possible settlements of the Standard litigation with Secretary of the Interior James R. Garfield. As a result of these preliminary talks, Goff delivered a formal proposal to the Justice Department on September 29. Standard directors offered to submit to any agreement worked out by Goff and Garfield. The oil trust would open its books to federal investigators. It would even contribute up to $100,000 to finance the investigation. And Attorney General Bonaparte would serve as the ultimate judge of the propriety of any final settlement. In return, the government would terminate the present antitrust litigation. Garfield considered this "a really astonishing proposal," adding that "Goff had done really well to induce them to make this offer." But other members of the administration wanted time to consider the matter.[92]

Both Kellogg and Morrison were firmly against compromise. On October 6, 1907, Kellogg wrote to the attorney general, declaring that he did not think that "the matters involved in this suit are proper subject for arbitration. They should be settled by the orderly procedure in courts regularly constituted for the purpose."[93] On October 22, Morrison sent a letter to Bonaparte also strongly advising against compromise. In an attempt to break this resistance, Standard Oil made another proposal. Senator Jonathan Bourne, Jr., of Oregon informed Roosevelt that if the government accepted Standard's offer, the oil combination would help him win renomination in 1908. Bourne's tactless statement killed the negotiations. Garfield labeled the effort "stupidly corrupt." And on October 25, Bonaparte informed the Standard directors that the president was constitutionally obliged to continue the prosecution.[94]

The Standard attempt to negotiate a settlement had been prompted by the desire to prevent criminal prosecutions. To achieve that goal the directors were ready to alter the structure and operating procedure of the oil combination. Although the Roosevelt administration refused to negotiate with Standard, it also failed to begin criminal prosecutions against the oil barons. Criminal prosecutions under the Sherman Act were difficult to win, but that consideration did not prevent Moody and Bonaparte from initiating a total of twenty-four such actions against antitrust offenders. On one occasion, Roosevelt called the Standard directors "the biggest criminals in the country."[95] Although he knew that they greatly feared criminal prosecutions and that a criminal suit could expedite the slow-moving civil suit, he did not order criminal prosecutions. Roosevelt's lack of action demonstrates his strictly limited commitment to antitrust enforcement.

Just as the Roosevelt administration dropped the idea of a criminal suit against Standard Oil, the New York financial community plunged into crisis. The prices of stocks and other securities fell steadily throughout 1907. By October many large financial institutions were having difficulty covering their obligations. On October 23 the Knickerbocker Trust Company folded, and the resulting withdrawal of funds from the leading financial institutions reached panic proportions. J. P. Morgan led a group of New York bankers in a frantic effort to prevent financial collapse. In mid-November George W. Perkins and Elbert H. Gary, key figures in the Morgan organization, asked Kellogg to postpone the Standard investigation until after the first of the year. Moreover, Morgan's representatives wanted the government to announce publicly the suspension of testimony: "We need all the help we can get from any and every direction to quiet things down."[96]

On November 23, 1907, Kellogg relayed the request to Attorney General Bonaparte, who three days later emphatically ruled out any public announcement regarding the oil investigation: "In my judgment, we can assent to no action which involves the admission that the course of the Administration in enforcing the Anti-Trust laws was, or is now, responsible for the existing financial complications. Moreover, an 'announcement,' such as Mr. Perkins desires, would be misconstrued and, indeed, would be useless for his purpose unless it was misconstrued." As for the actual matter of postponement, Bonaparte observed, "I do not know that thirty days delay would make any appreciable difference."[97] His position represented a marked retreat from his earlier concern with bringing the case to a speedy conclusion. Apparently the panic had shaken his resolve to press forward with the case.

On December 2, 1907, the Standard attorneys formally requested a postponement of testimony until January 6, and Kellogg promptly granted the request. The special prosecutor, however, insisted that evidence from several earlier state antitrust cases and federal investigations be read into the record as evidence during the postponement. This arrangement kept the proceedings moving forward in some fashion and prevented the public appearance of delay. Kellogg admitted that he was "loath to adjourn this case at this time" but thought that the postponement a "good policy." Bonaparte agreed. On December 3 he approved Kellogg's arrangement and termed it "very judicious."[98] The Roosevelt administration backed away once again from energetic pursuit of the case against Standard Oil.

A 1908 Standard Oil attempt to settle a huge rebating fine provides still another example of the administration's basic contempt for the antitrust process. In August 1907 a federal circuit court judge, Kenesaw Mountain Landis, ruled that Standard Oil of Indiana had violated the Elkins Act by accepting rebates on oil shipments from its Whiting, Indiana, refinery to St. Louis. He imposed the maximum fine of $20,000 for each of the 1,462 counts against the company, arriving at a final figure of $29.24 million.[99] In April 1908 John D. Archbold contacted Attorney General Bonaparte and proposed a compromise of both the $29 million fine and the antitrust suit. He suggested that the company concede its guilt and pay fines on "five per cent of the counts contained in the indictments against them" Archbold calculated that these fines would total $508,950. In return the government would drop both suits. Although he rejected this offer, Bonaparte kept the negotiations open by asking for time to develop a counterproposal.[100]

In order to obtain more information on the rebate case, Bonaparte contacted the Bureau of Corporations. The bureau labeled the oil trust a "habitual offender" and claimed possession of evidence against the combination justifying a fine ranging from $9,764,000 to $195,280,000.[101] In mid-May Bonaparte presented the government's offer. He could not compromise the Landis decision for any amount—the $29 million fine would stand—but he would settle the antitrust suit. Standard would have to plead guilty to one count in five in each of the indictments and it would have to pay fines totaling $1,744,000. The government then would agree to dismiss the remaining counts in order to avoid "the trouble, delay, expense and uncertainty involved." Through Senator Bourne, Archbold indicated that he needed time to consider this offer, and Bonaparte agreed that Archbold could wait until June 5.[102]

On June 4, Archbold accepted Bonaparte's proposal, but he imposed some strict conditions. The administration would have to halt the antitrust investigation and grant immunity from similar prosecutions in the future. Then the oil trust could negotiate with the government in a spirit of "harmony."[103] Archbold's conditional acceptance of Bonaparte's offer was geared to enhancing the possibility of a future settlement of the $29 million rebate fine. If Bonaparte agreed to Archbold's terms, then Standard's harmonious relations with the government might provide a springboard for a fine settlement. If Bonaparte declined, then further negotiations would result—opening the way for a possible settlement of the Landis fine.

Bonaparte could not promise future immunity to the oil trust; hence further negotiations were necessary. The government insisted on a reorganization of the trust before immunity could be considered, and conferences on a reorganization plan occurred during late June and early July. But on July 20 Archbold suddenly broke off all negotiations. Two days later, Roosevelt and Bonaparte discovered why: circuit court Judge Peter S. Grosscup overturned the $29 million fine. Roosevelt explained: "Archbold must have known in advance what the Grosscup decision was to be." This suspicion was reinforced on July 27, when T. Parmalee Prentice, John D. Rockefeller's son-in-law, gave a party in honor of Grosscup.[104]

The negotiations between April and July 1908 again underscored the Roosevelt administration's lack of faith in the Sherman Act. Bonaparte reflected his boss's frequently stated attitude toward antitrust litigation when he offered to compromise the suit to avoid "the trouble, delay, expense and uncertainty involved." Antitrust litigation was expensive and uncertain, but federal prosecutors already had been gathering evidence against Standard Oil for two years. Moreover state investigations, the Bureau of Corporations, and the Interstate Commerce Commission had amassed a vast amount of highly damaging evidence. Roosevelt had made the Standard combination an important political issue. A healthy oil industry was vital to the national economy. In short, if Roosevelt had been committed to strict antitrust enforcement, this was the one case he should have been unwilling to compromise.

Archbold's actions in the course of these negotiations proved that the Standard leadership was not overly concerned with the antitrust suit either. From the beginning Archbold concentrated on disposing of the $29 million fine; the antitrust suit was secondary. As soon as Archbold learned that Grosscup was about to overturn the Landis fine, he promptly rejected Bonaparte's offer to drop the antitrust suit. Apparently both the government and Standard Oil were aware that the uncertainties of antitrust litigation offered the oil trust innumerable opportunities to avoid any significant disruption of normal operations.

Roosevelt's indifferent pursuit of the antitrust case did not diminish his enthusiasm for exploiting Standard's poor public image. To obscure the fact that his administration was easing its policy toward the trusts in the face of the panic of 1907 and to promote a series of new federal controls over business, Roosevelt attacked Standard Oil and several other combinations in a special message to Congress in January 1908. The president claimed that "the speculative folly and flagrant dishonesty of a few great men of wealth" had produced the panic. The Standard Oil leadership was particularly offensive: "Every measure for honesty in business that has been passed in the last six years has been opposed by these men . . . with every resource that bitter and unscrupulous craft could suggest and the command of almost unlimited money secure."[105]

Roosevelt also injected the Standard Oil issue into the 1908 presidential election. Roosevelt first advised Taft to refuse any Standard campaign contributions and to emphasize the government's energetic pursuit of the antitrust case. The president then tried to link Standard Oil to the presidential campaign of the Democratic nominee, William Jennings Bryan. On September 21 Roosevelt publicly noted that Charles N. Haskell, governor of Oklahoma and treasurer of the Democratic party, had close Standard Oil ties. In the summer of 1908 Haskell had used his influence as governor to dissolve an injunction preventing the oil combination from constructing a pipeline in Oklahoma. Despite this action, Haskell retained his position of responsibility in the Democratic party. Roosevelt urged Taft to emphasize Haskell's Standard Oil connection in the campaign, and the consequent pressure resulted in Haskell's resignation.[106]

Ironically John D. Rockefeller undid most of Roosevelt's careful work. On October 29, 1908, he announced that he would vote for Taft. Roosevelt was furious. He exclaimed that Rockefeller's proclamation was "a perfectly palpable and obvious trick." Taft promptly announced that he did not want Standard Oil's support, adding that "if the Standard Oil were anxious to bring about my election, the last thing they would have done would have been to advertise their support for me."[107]

By the end of his presidency, Roosevelt had developed a policy toward Standard Oil based on calculated deception. He adamantly refused to act on the numerous complaints against the oil trust during his first five years in the White House. He consistently asserted that industrial combination was inevitable and that antitrust litigation was futile. He brought suit against Standard Oil only after his exploitation of the Bureau of Corporations' *Report on the Transportation of Petroleum* made antitrust action politically profitable. He refused to initiate criminal suits against Standard leaders even though such suits would have expedited the civil case. His attorney general suspended the Standard Oil investigation during the panic of 1907. And the president would have compromised the case in 1908 if Archbold had not broken off negotiations. On the other hand, Roosevelt seized every opportunity to portray himself as the enemy of the oil trust. He was the first American president to recognize that the woefully ineffective antitrust process was an excellent tool for maximizing the public appearance of action against big business while minimizing the economic disruption that frequently accompanied government intervention.

(6)

The Rule of Reason

William Howard Taft became president of the United States on March 4, 1909. Like his predecessor, he held deep convictions concerning the right of private property and the unequaled virtues of the capitalist system. In 1906, Taft had solemnly warned Yale students—a group hardly noted for its revolutionary zeal—about seductive reformers, "who yearn for an entirely different system and radical change, in which men are to be governed solely by love and not by motive of gain." Taft also heartily endorsed the trend toward industrial combination. In a special message to Congress on January 7, 1910, he declared that combination "has become as essential in modern progress as the change from the hand to the machine."[1]

In marked contrast to Roosevelt, Taft respected the law at least as much as he respected property rights. In fact, he viewed the law as the best guarantee of stability and order, so essential to the preservation of property. He believed that an efficient and equitable judicial system could defuse many of the explosive issues created by the emerging industrial order. Sometimes his expressions of devotion to the law bordered on the absurd. In 1911, he told the Pocatello Chamber of Commerce: "I love judges, and I love courts. They are my ideals, that typify on earth what we shall meet hereafter in heaven under a just God."[2] Taft's veneration of the law amazed presidential assistant Archie Butt:

President Taft will do anything if he has a law on which to base his act. The law to President Taft is the same support as some zealots get from religious faith. And the fact that the law is unpopular would not cause him to hesitate a minute.[3]

Taft believed that the Sherman Act was clear, and he had little patience with businessmen who complained about the law's vagueness. In a letter to Otto Bannard, a New York banker, Taft exploded: "The suits under the Sherman act are not speculative, and I can tell without the slightest difficulty whether a company is violating the law, and so can anybody else who wants to be genuine about it and look into the matter." But in the same letter, Taft revealed his own confusion about the act. "Of course," he wrote, "there must be some doubt as to the intent and purpose of particular combinations."[4] This doubt is precisely what troubled businessmen.

Because he erroneously considered the Sherman Act clear and because he respected all laws, Taft committed his administration to strict antitrust enforcement. In his second annual message to Congress on December 6, 1910, Taft announced that he was not going to follow Roosevelt in pressing for elaborate new federal regulatory powers over big business: "It seems to me that the existing legislation with reference to the regulation of corporations and the restraint of their business has reached the point where we can stop for a while and witness the effect of the vigorous execution of the laws on the statute books."[5] Taft lived up to this promise. His four-year administration initiated a record number of antitrust cases—ninety—compared to the Roosevelt administration's forty-four in seven years.[6]

George W. Wickersham, Taft's attorney general, was responsible for antitrust enforcement. Wickersham, who shared most of his boss's basic attitudes toward big business and the law, and believed that large-scale enterprise was an economic necessity and that "size alone does not constitute monopoly."[7] These views reflected Wickersham's close ties to big business. The new attorney general had worked as a top corporate lawyer for over twenty-five years and reportedly left a $100,000 per year practice to join Taft's cabinet. Shortly after Wickersham took office, Joseph H. Choate, the distinguished lawyer who helped Standard Oil avoid dissolution in 1892, expressed high expectations for the new attorney general at a dinner of the Bar of the City of New York: "He has been a corporation lawyer, a defender of institutions which twelve months ago were everywhere condemned. It is quite time that they had their innings."[8]

Wickersham understood the political necessity of dealing with the problem of monopoly. He also shared Taft's belief that the Sherman Act clearly and effectively dealt with the problem. On April 30, 1909, the new attorney general declared, "The principles underlying the law are assuredly now understood, and any attempt at this time . . . to combine in a form of a trust . . . would evidence such a deliberate intention to break the law as to compel the government to use all or any of the remedies given by the law."[9] Some businessmen felt betrayed when they realized that Wickersham intended to push antitrust litigation. In 1911 the *Wall Street Journal* bitterly remarked:

Corporate baiting is to him good politics. . . . There will come a day of reckoning when the man who got away with the vote will be gone after with a shotgun, because his disintegration of "big Business" has resulted in increasing the cost of shoes and plows.[10]

The *Journal's* reaction should not obscure the essentially conservative intent behind Wickersham's policy. Both Taft and his attorney general were strong defenders of the status quo, who believed that vigorous and equitable law enforcement could ameliorate pressing social problems and prevent radical economic reform. This conviction led them to impart a false clarity to the notoriously vague Sherman Act, and it prompted them to press antitrust litigation against their natural political allies on Wall Street—including Standard Oil.

Vigorous prosecution of the case against the oil trust was politically, as well as philosophically, necessary for the Taft administration. Taft had joined Roosevelt in making his opposition to Standard Oil an issue in the 1908 presidential campaign. After taking office, Taft continued the policy of exploiting public hostility toward Standard Oil. In December 1909, Secretary of War Jacob M. Dickenson, on Wickersham's advice, announced that the War Department was immediately halting all purchases of petroleum products from the Standard Oil Company or any of its affiliates.[11]

In March 1910 a bill establishing the Rockefeller Foundation as a national corporation, which would direct the income from $100 million toward various charitable projects, was introduced in Congress. Despite the humanitarian intent of the bill, Wickersham firmly opposed its passage, labeling it "an indefinite scheme for perpetuating vast wealth." Taft was also alert to the necessity of avoiding any federal support for Rockefeller, regardless of the circumstances. He wrote, "I agree with your . . . characterization of proposed act to incorporate John D. Rockefeller." Sponsors withdrew the bill in the face of administration opposition. They reintroduced it in 1912 in a modified form, but its subsequent failure led to the incorporation of the Rockefeller Foundation in New York State.[12]

In April 1911, Archie Butt reported another incident that reveals Taft's strenuous efforts to avoid any association with the oil trust.

At three there was to be a reception of the Children of the Revolution. I had already directed one of the junior ads to be there to present them to the President, when I got a message from him that he wanted me and no one else. I was somewhat mystified until I learned that ex-Senator Aldrich, with his daughter and her husband, Mr. Rockefeller, were to be at lunch and that nothing was to be said about them being there. It is strange how men in public office shudder at the names of Aldrich and Rockefeller.

All guests for luncheon entered the White House by the main entrance, but the President gave the tip that he wanted them to come in unobserved, so the usher telephoned Mr. Aldrich to drive to the East entrance where he would meet them and take their wraps. Of course, the Senator understood. The President said he wanted no note of the fact that they were there entered on any of the books or made known even at his office.[13]

The Standard Oil case had achieved considerable momentum of its own by the time Taft entered the White House. Kellogg and Morrison concluded their lengthy investigation in February 1909 and immediately began drafting the prosecution briefs and the proposed dissolution decree. Attorney General Wickersham took office just in time to help put the finishing touches on those documents, and the federal prosecutors began presenting their case to the St. Louis circuit court on April 5, 1909, only one month after Taft's inauguration. Walter H. Sanborn of Minnesota served as presiding judge. The other members of the court were Elmer B.

Adams of Missouri, William C. Hook of Kansas, and Willis Van Devanter of Wyoming, a future Justice of the United States Supreme Court. From the government's standpoint, the most important fact about this court was that Sanborn and Adams both had voted to dissolve the Northern Securities Company in 1903. The other two judges were not on the court at that time.[14]

The government prosecutors argued that the Standard combination had achieved its domination of the petroleum industry through the Jersey Standard holding company, which had controlled the stock of firms in all phases of the industry since 1899. It was therefore in violation of section 1 of the Sherman Antitrust Act, which declared illegal "every contract, combination in the form of a trust or otherwise, or conspiracy, in restraint of trade or commerce among the several States, or with foreign nations." The prosecutors buttressed this point by reminding the judges that the United States Supreme Court had ruled against the Northern Securities holding company in 1904. In that case, Justice John Marshall Harlan held that stock transfers to form a holding company were illegal if they "directly or necessarily" caused a restraint of trade. Given the magnitude of the oil combination's interstate business, an application of Harlan's rule to the Jersey Standard holding company all but guaranteed conviction.[15]

Kellogg and Morrison also argued that the Jersey Standard holding company violated section 2 of the Sherman Act, which declared illegal any attempt "to monopolize any part of the trade or commerce among the several States, or with foreign nations." They contended that the oil combination had used its vast power to drive out competitors through unethical practices, including rebating, espionage, price cutting, and the use of bogus independents. Throughout the hearings, the government repeatedly documented Standard's use of these methods. The federal attorneys argued that "a monopoly exists where a great aggregation of capital, like the Standard Oil, is coupled with oppressive use of the power such wealth gives."[16]

The aspiring trustbusters concluded their presentation by noting the broad power of the court under the Sherman Act to restore competition to the petroleum industry and then requesting that the court issue a decree enjoining Jersey Standard "from exercising any control over defendant corporations by stock ownership or otherwise."[17] By preventing Jersey Standard from performing functions such as voting the stock of and receiving dividends from its subsidiaries, the government proposed to dissolve the oil combination into several independent and competing firms.

The government plan divided the combination along corporate, rather than functional, lines. Because most of the subsidiary companies specialized in a certain phase of the industry, they were incapable of independent operation and had to cooperate closely with the other Standard companies in order to stay in business. Table 1 demonstrates the lack of integration of almost all the Standard subsidiaries. For example, the three principal Standard companies in Ohio were Ohio Oil, which produced crude, Buckeye Pipeline, which operated pipeline and storage facilities, and Standard Oil of Ohio, which refined and marketed petroleum products. These companies together controlled over 90 percent of the Ohio oil industry and could not operate independently. Ohio Oil would have to transport its crude through the Buckeye Pipeline network to the Ohio Standard refineries in order to remain in business. The proposed decree also failed to disturb the combination's nationwide system of marketing territories that guaranteed only one Standard affiliate would market petroleum products in a given region.

Kellogg and Morrison knew that the court possessed adequate legal power to achieve a real dissolution, yet they failed to formulate a plan to guarantee actual competition. Such a plan could have been developed by grouping certain Standard companies in order to form integrated and potentially competitive units. Their failure to devise such a proposal suggests that they were more interested in winning a politically important case than in restoring competition to the industry.

The major problem for the Standard Oil defense team, headed by John G. Milburn, was to overcome the difficulties presented by the Northern Securities precedent. The defense first attempted to exploit an argument suggested by Supreme Court Justice Oliver Wendell Holmes in a dissenting opinion in the *Northern Securities* case. Holmes had asserted that the language of the Sherman Act did not "require all existing competitions to be kept on foot, and . . . invalidate the continuance of old contracts by which

TABLE 1
FUNCTIONAL ORGANIZATION OF STANDARD NEW JERSEY
AND DISAFFILIATED COMPANIES AFTER DISSOLUTION

PRIMARY FUNCTIONS

COMPANY	Producing	Refining	Domestic Marketing	Foreign Marketing	Domestic Manuf. and/or Distrib. of Specialty Products	Pipeline Transportation and Storage	Tanker Transportation
Standard New Jersey	x	x	x	x	x	x	x
Anglo-American Oil				x			x
Atlantic Refining		x	x				
Borne, Scrymser				x	x		
Buckeye Pipe Line						x	
Chesebrough Manufacturing				x	x		
Colonial Oil				x			
Continental Oil			x				
Crescent Pipe Line						x	
Cumberland Pipe Line						x	
Eureka Pipe Line						x	
Galena-Signal Oil			x	x	x		
Indiana Pipe Line						x	
National Transit						x	
New York Transit						x	
Northern Pipe Line						x	
Ohio Oil	x					x	
Prairie Oil and Gas	x					x	
Solar Refining			x				

	1	2	3	4	5	6	7
Southern Pipe Line						x	
South Penn Oil	x						
South-West Pennsylvania Pipe						x	
Standard California	x	x	x			x	x
Standard Indiana		x	x				
Standard Kansas		x					
Standard Kentucky			x				
Standard Nebraska			x				
Standard New York		x	x	x			x
Standard Ohio		x	x				
Swan and Finch			x		x		
Union Tank Line					x	x	
Vacuum Oil		x	x	x	x		
Washington	x						
Waters-Pierce		x	x	x			

SOURCE: George S. Gibb and Evelyn H. Knowlton, *The Resurgent Years: 1911–1927* (New York: Harper and Brothers, 1956), pp. 8-9.

former competitors united in the past.''[18] Standard lawyers therefore emphasized that the defendant corporations had not competed with each other since the first Standard trust agreement in 1879—fully thirty years ago. Moreover, that agreement, and all the oil trust's subsequent organizational arrangements, resulted from the natural course of economic development in the petroleum industry during the last half century.[19] Their argument had two serious weaknesses: it was based on a minor point in a dissenting opinion, and the "naturalness" of the Standard combination was open to serious question.

The Standard attorneys also tried to distinguish the Jersey Standard holding company from the illegal Northern Securities holding company by noting that the latter was a railroad combination and thus was more directly involved in interstate commerce than Standard Oil was. Here the defense attempted to exploit the fact that many of the important Supreme Court decisions supporting government prosecutions under the Sherman Act—the *Trans-Missouri, Joint Traffic,* and the *Northern Securities* cases—had involved railroad combinations. The lawyers wanted the court to view Standard Oil as a manufacturing concern, which was exempt from the Sherman Act under the *Knight* rule.[20] This argument, however, had a significant limitation. It failed to confront the fact that the oil trust was deeply involved in the interstate transport of crude and refined petroleum both by rail and by company-owned pipelines. Standard critics had consistently emphasized the importance of the transportation sector of the industry, citing such abuses as rebates, drawbacks, and the failure of Standard pipelines to act as common carriers. Furthermore, the government prosecutors had recounted these misdeeds in great detail in their briefs. Hence there was little chance that the court could be convinced that the oil trust was not directly involved in interstate commerce. The *Northern Securities* precedent and the oil trust's long history of anticompetitive practices prevented the defense team from building a strong case.

On November 20, 1909, eight months after the presentation of arguments, the circuit court handed down its opinion. Presiding Judge Walter Henry Sanborn, for a unanimous court, ruled in favor of the government. The court found the *Northern Securities* precedent controlling on the issue of a section 1 violation. Sanborn noted that many of the companies controlled by Jersey Standard were potentially competitive, yet the stock transfers to the holding company gave Jersey Standard "the absolute power to prevent competition among any of these corporations." Therefore the "necessary effect" of the stock transfers "was under the decision in the case of the Northern Securities Company, a direct and substantial restriction of [interstate] commerce."[21]

The court also ruled that Standard Oil had attempted to monopolize the petroleum industry in violation of Section 2 of the Sherman Act, but Sanborn's opinion did not go into the matter of anticompetitive practices. Rather he held that "the combination and conspiracy in restraint of trade and its continued execution, which have been found to exist, constitute illegal means by which the conspiring defendants combined, and still combine and conspire to monopolize a part of interstate and international commerce.[22] In other words, the court ruled that when Standard Oil combined in violation of section 1, and thereby dominated interstate commerce in the oil industry, it automatically violated section 2. This argument allowed the court to avoid ruling on the overwhelming mass of evidence regarding Standard's business practices. The four judges were ill equipped to deal with the complex economics of this case,[23] and were reluctant to evaluate matters beyond their competence. Although they had confronted cases dealing with economic issues before, none had matched the size and complexity of this one.

After ruling against the oil trust, the court followed the prosecution's request and decreed that Jersey Standard was prohibited from "voting any of the stock in any of the subsidiary companies" and from "declaring or paying any dividends to the Standard company on account of any stock of the subsidiary company held by the Standard company."[24] This decree would do little to restore actual competition to the petroleum industry. The court, however, merely followed the prosecution's proposal. The judges were not in a strong position to grasp the weakness of the government plan.

The court, without prosecution prompting, inserted another provision into the decree of utmost importance to the future of the Standard organization. It declared that "the defendants are not prohibited by this decree from distributing ratably to the share-holders of the principal company the shares to which they are equitably entitled in the stocks of the defendant corporation that are parties to the combination."[25] This qualification meant that each Jersey Standard stockholder would receive a fractional share of stock in every subsidiary. In each case the denominator of the fractional share would be the total number of Jersey Standard shares—983,383—and the numerator would be the number of shares held by Jersey Standard in a given subsidiary. For example, the holder of one share of Jersey Standard stock would receive 599,994/983,383 of a share of Ohio Oil stock and 2,747/983,383 of a share of Waters-Pierce stock. Because eight individuals, or their estates, controlled over 50 percent of all Jersey Standard stock, the stock distribution plan guaranteed that this small group would own a controlling interest in every one of the subsidiaries. Therefore the prospect for real competition among these companies was negligible.[26]

The court was fully aware that the pro-rata distribution of stock opened the door to continued combination—Standard trustees had employed this method of stock distribution to avoid dissolution following the 1892 Ohio Supreme Court decision against the oil trust, and the government briefs had covered that process in considerable detail. In fact, Sanborn recalled these events in his opinion, noting that "this method of distribution appears to have deterred" dissolution.[27] In a vain attempt to compensate for this transparently defective stock distribution plan, the court appended two prohibitions to the decree. It "enjoined and prohibited" any further combination in restraint of trade through "the use of liquidation certificates" or through "any express or implied agreement or arrangement."[28] But these prohibitions were of little value. The oil barons could maintain their stock control without the use of liquidation certificates. And the sweeping prohibition against any "express or implied agreement" was so broad that it was useless.

If the court had been sincerely interested in restoring competition to the industry, it could have required Jersey Standard to assign its stockholders blocks of full shares in one or more companies. That way different companies would have had different owners, and the incentive to compete would have been restored. This method of stock distribution had the added benefit of allowing full shares, or blocks of shares, to small stockholders, who under the pro-rated distribution plan received a collection over thirty absurdly small fractions.

Sanborn offered no explanation of why the circuit court allowed a method of stock distribution that could lead easily to the frustration of the intent of the decree, but the structure and the language of his opinion suggest two possible explanations. First, Sanborn had repeatedly demonstrated his reluctance to deal with the complex economics of the case. He refused to rule on the matter of Standard anticompetitive practices, and he accepted the government's plan for a corporate, rather than functional, dissolution of the trust. By following the pro-rata stock distribution scheme worked out by Standard's own attorneys in 1892, the court avoided the formidable task of remodeling the Standard stock structure in a way that was both equitable and conducive to competition. Second, the language of the opinion indicates that Sanborn and the other judges were concerned that they avoid damage to the property rights of the defendants and to the petroleum industry, which was very important to the national economy. Sanborn described his task of fashioning a decree as "grave and delicate." He emphasized his "duty not to deprive the defendants of their right to engage in lawful competition for interstate and international commerce."[29] Quite possibly, the judges felt that any reorganization of stockholding patterns violated property rights, and they refused to do that, even if the cost was continued monopoly.

Kellogg was elated with the court's decision. He was even pleased with the decree. On the day the decision was handed down, he sent a telegram to Wickersham declaring: "Decision in Standard Oil case is complete victory for Government . . . opinion and decree are all that Government asked."[30] Kellogg took this position despite the glaring inadequacies of the decree for several reasons. One was an obvious political stake in placing the best possible interpretation on his work. But more importantly, Kellogg was a conservative corporation lawyer who believed that antitrust litigation was a useful method of curbing some of the most glaring industrial abuses and defusing social tensions, which might lead to genuine radicalism. Therefore, a well-publicized court decision against a notorious corporation like Standard Oil, enforced by a weak decree, which posed no real threat to property interests, suited both his ideological and political purposes.

Attorney General Wickersham and President Taft also welcomed the decision and approved of the decree. They apparently considered the decision a genuine setback for the Standard combination and another successful episode in their continuing policy of maximum exploitation of the oil trust. Wickersham told the *New York Times,* "It's a great victory. I consider this one of the most important decisions ever rendered in this country."[31] Taft expressed his pleasure to Kellogg: "I congratulate you on . . . the complete victory you have won . . . much of which is due to the thorough preparation and presentation, on your part of the government's case." But the president also needlessly worried about the possibly disruptive effect of the decision on the business community.[32]

While the Taft administration celebrated its victory, Standard attorneys appealed the case to the United States Supreme Court. The nine justices of the high court were split into three fairly distinct groups on the specific issue of the legality of a holding company. Four could be expected to rule against Standard Oil of New Jersey. John Marshall Harlan was the most consistent supporter of strict antitrust enforcement on the court. He had been the lone dissenter in *Knight;* he had supported the government in both the *Trans-Missouri* and *Joint Traffic* decisions; and, most important, he had written the majority opinion on the *Northern Securities* case. Joseph McKenna and William R. Day also voted with the government in the *Northern Securities* decision and thus would be predisposed to rule against Standard Oil. Willis Van Devanter, who had already ruled against the oil trust while serving on the St. Louis circuit court, was now on the Supreme Court and would almost certainly again vote against Standard. In fact, the Taft administration had placed Van Devanter on the Supreme Court partly because of his earlier vote against the oil combination.[33]

Standard attorneys could anticipate a clearly sympathetic hearing from only two members of the court. Chief Justice Edward D. White, a consistent opponent of effective antitrust enforcement, had voted against the government in the *E.C. Knight* and *Joint*

Traffic cases and had written dissenting opinions in the *Trans-Missouri* and *Northern Securities* decisions. Oliver Wendell Holmes, the most accomplished jurist on the court, also had written a dissenting opinion in the *Northern Securities* case. In 1910 Holmes wrote, "I don't disguise my belief that the Sherman Act is humbug based on economic ignorance and incompetence."[34] The three remaining justices did not have a clear public record on the antitrust issue. None of them had participated in the critical *Northern Securities* decision of 1904. Charles Evans Hughes and Joseph R. Lamar had joined the court in 1910, and Horace H. Lurton had been placed on the high tribunal a year earlier.[35] If the defense attorneys could convince White and Harlan of the legality of the Jersey Standard holding company, perhaps these two influential jurists could rally a majority to the oil combination's defense.

The Supreme Court heard arguments in the *Standard Oil* case between January 12 and 17, 1911. Attorney General Wickersham joined Kellogg in presenting the case and participated in developing the government's strategy. The prosecutors decided to use the same approach that already had proved successful before the circuit court. Wickersham and Kellogg again argued that Standard dominated the industry by means of the Jersey Standard holding company. They again reviewed the oil trust's long history of anticompetitive practices. Standard's dominant position in the industry and its methods indicated restraint of trade and monopolization in violation of sections 1 and 2 of the Sherman Act.[36] Significantly, neither Wickersham nor Kellogg made an effort to alter the toothless decree. After working on the case for eighteen months, Wickersham must have understood the inadequacies of a simple pro-rata stock distribution. He, like Kellogg, was more interested in a well-publicized victory than actual dissolution of the oil trust.

Since the Standard Oil attorneys had based their presentation before the circuit court on arguments suggested by White and Holmes in the *Northern Securities* case, they also repeated the same basic contentions before the Supreme Court. The defense argued that the Sherman Act did not apply to long-established combinations resulting from the normal course of business. They then portrayed the oil combination as a "natural outgrowth and development of the business organized in 1862 by the Rockefellers." Standard lawyers also asserted that the Sherman Act could not constitutionally reach the oil combination because it primarily engaged in production, which, under the *E. C. Knight* rule, was beyond the federal commerce power. In addition, the defense claimed that the circuit court decree had impaired property rights and restricted liberty of contract in violation of the due process clause of the Fifth Amendment.[37] Although these arguments suffered from several serious weaknesses, they were the best that the Standard attorneys could devise.

Following the conclusion of final arguments, the financial elite in New York and the political elite in Washington anxiously awaited the Supreme Court's decision. During April and May rumors repeatedly circulated that the court was about to announce its ruling. Each Monday, the day the tribunal delivered its decisions, the tiny Supreme Court chamber in the Capitol building was filled to capacity. Kellogg, who was vacationing in London, wired each week inquiring about news of a ruling. On May 15, one of the largest crowds of the spring gathered in anticipation of the Standard Oil decision. At 4:00 P.M., after a long and tedious day of decisions, the Chief Justice routinely declared, "I have also to announce the opinion of the Court in No. 398, the United States against the Standard Oil Company."[38] There was a brief moment of disorder in the crowded courtroom as several newsmen dashed for telephones, and the rest of the crowd strained to hear the Chief Justice. White held the printed text of the decision before him, but he related its contents from memory. According to one observer, "Sometimes he could be heard by the intent listeners, and sometimes he could not. He did not seem greatly to care whether or not his words were audible."[39]

Chief Justice White's twenty-thousand-word opinion upheld the circuit court's conviction of Standard Oil. White followed the lower court in avoiding the massive array of economic evidence concerning rebates, espionage, and other related matters. He noted that an adequate analysis of the facts presented by attorneys on both sides would be "a duty difficult to rightly perform and, even if satisfactorily accomplished, almost impossible to state with reasonable regard to brevity."[40] Like the circuit court, the Chief Justice argued that the creation of the holding company in 1899 violated sections 1 and 2 of the Sherman Act because that form of corporate organization was not created "as a result of normal methods of industrial development, but by new means of combination which were resorted to in order that greater power might be added than would otherwise have arisen had normal methods been followed."[41] Jersey Standard's heavy-handed exercise of power "fortified" this conclusion, White added. The other eight members of the Court unanimously supported the Chief Justice on these points. The mass of evidence compiled by the government against the oil trust was too conclusive to permit any other decision.

Although White found the oil combination in violation of the Sherman Act, he considered the innocuous circuit court dissolution plan too harsh and modified three aspects of the decree. He extended the time limit for compliance from thirty days to six months, overturned a provision that enjoined all Standard companies from engaging in interstate commerce until the dissolution was complete, and, most important, interpreted a provision in the decree that prohibited Standard affiliates from entering into restrictive agreements with each other after the dissolution as permitting "normal and lawful contracts or agreements."[42] Given the nonintegrated nature of the Standard affiliates and existing business patterns, extensive informal cooperation after the dissolution was inevitable. By failing to reorganize the Standard companies into coherent, potentially competitive units and by specifically legalizing trade agreements between ex-affiliates, White went a long way toward guaranteeing a de facto continuation of the oil trust.

But the further enfeeblement of the decree was overshadowed by another, totally unexpected, aspect of the decision. White inserted into his Standard Oil opinion an interpretation of the Sherman Act that he had advocated unsuccessfully for fifteen years: the rule of reason.

The rule of reason modified the sweeping provisions of the Sherman Act, which declared illegal every contract or combination in restraint of trade, to prohibit only restraints that were unreasonable or against the public interest. Under the rule, judges determined what contracts and combinations were reasonable and what constituted the public interest according to the circumstances of the case at hand. This expansion of judicial discretion allowed the courts to declare legal monopolies that they considered socially beneficial.

The rule of reason originated in the branch of the common law dealing with contracts in restraint of trade. Most authorities agree that early common-law judges were overwhelmingly hostile toward such contracts. The seventeenth century, however, witnessed a gradual erosion of certain legal prohibitions against these restraints. A series of court decisions established the principle that restraints were legal if they were ancillary to a legitimate business transaction, such as the sale of a business or an employment contract; partial (limited in time and space); and reasonable in the light of the interests of the contracting parties and the general public. The third provision of this formula was the original application of the rule of reason.[43]

During the nineteenth century, the concept of partial restraints faded from the law of ancillary contracts in both Britain and America. The 1894 case of *Nordenfelt* v. *Maxim Nordenfelt Company* marked the final demise of the concept.[44] In that decision, the British House of Lords upheld a sales contract that prohibited the seller of an arms manufacturing firm from reentering that business anywhere in the world for a period of twenty-five years. The Lords ruled that arms manufacturing necessarily entailed a worldwide market because governments were the most important customers and that the prohibition was reasonable. By 1900 the test of reasonableness had become the controlling principle in cases concerning restraints ancillary to a larger transaction on both sides of the Atlantic.

Nineteenth-century British and Commonwealth courts also applied the rule of reason to agreements and loose combinations among competitors, including arrangements aimed at price fixing, profit pooling, output limitation, and territorial devision. These agreements were not ancillary to a main contract of sale or employment; therefore they represented a distinct break with the traditional concept of the rule. Business consequently gained greater freedom of contract. The 1879 case of *Collins* v. *Locke* illustrates this trend. The court upheld an agreement to distribute work and profits between four stevedoring firms, even though the judges recognized that the arrangement attempted to prevent competition and keep prices high. They considered the contract valid "if carried into effect by proper means, that is, by provisions reasonably necessary for the purpose, though the effect of them might be to create a restraint upon the powers of the parties to exercise their trade."[45]

In the United States the courts generally did not apply the rule of reason to non-ancillary restraints. American judges therefore usually looked with disfavor upon restrictive agreements among competitors. For example, the Ohio Supreme Court, in the 1880 case of *Central Ohio Salt Company* v. *Guthrie,* refused to enforce a pooling agreement among Ohio salt manufacturers that fixed prices, divided profits, and limited output: "The clear tendency of such an agreement is to establish a monopoly, and to destroy competition in trade, and for that reason, on the grounds of public policy, courts will not aid in its enforcement." The court then added that the reasonability of an agreement was beside the point: "It is no answer to say that competition in the salt trade was not in fact destroyed, or that the price of the commodity was not unreasonably advanced."[46]

Milton Handler, a noted authority on antitrust law, summarized the difference between the British and American decisions on restraints of trade that had developed by the 1890s:

The fact is that the English and some American courts regarded the rule of reason as applicable to horizontal arrangements between competitors, which curbed, controlled, or eliminated competition, and which were not ancillary to any major transaction. Many American courts, on the contrary, either explicitly regarded the rule of reason as inapplicable to such arrangements, or, without mention of the rule, branded the activities under consideration as unlawful.[47]

The United States Supreme Court first seriously considered the common-law background of the Sherman Act in *United States* v. *Trans-Missouri Freight Association* (1897), a case that involved a government attempt to dissolve a rate-fixing agreement among eighteen previously competing railroads in the Southwest. When the lower courts ruled the agreement valid, the government appealed the case. Before the Supreme Court, railroad attorneys argued that the Sherman Act must be interpreted in light of the common-law decisions in recent years that allowed reasonable restraints on trade. For obvious tactical reasons, the railroad lawyers chose to emphasize the trend of British rather than American decisions. Because the *Trans-Missouri* agreement brought stability to the ruinously competitive railroad industry, they argued that it was reasonable.[48]

A closely divided court ruled in favor of the government. Justice Rufus W. Peckham, speaking for a five-man majority, accepted the railroad attorney's argument that reasonable restraints were valid at common law without examining the precedents on restraint of trade. But Peckham claimed that the language of the Sherman Act, literally interpreted, went beyond the common law: "the plain and ordinary meaning of such language is not limited to that kind of contract which is in unreasonable restraint of trade, but all contracts are included in such language, and no exception or limitation can be added without placing in the act that which has been omitted by Congress."[49] Peckham added that the Court would be inviting chaos if it accepted the notion that rate fixing could be reasonable. After all, complex economic factors far beyond the competence of the average judge determined the reasonableness of a rate. Peckham also asserted that the Court did not have the proper authority to depart from the language of the Sherman Act.

Peckham's opinion proved that a profoundly conservative judge, who wrote some of the most reactionary decisions in the history of the court, could be a strong ally of the antimonopolists. He supported the effort to halt unrestrained monopoly because he feared the eventual destruction of small businesses and with them the laws of competition. A nineteenth-century rugged individualist, Peckham warned against the passing of the independent entrepreneur: "It is not for the real prosperity of any country that such changes should occur which result in transferring an independent business man, the head of his establishment, small though it might be, into a mere servant or agent of a corporation."[51] During the 1890s Peckham became one of the foremost judicial champions of effective antitrust laws.

Justice White, speaking for a four-man minority, vigorously dissented from Peckham's literal interpretation of the Sherman Act. White noted, but made no effort to prove, that the common law prohibited only unreasonable restraints on trade. He sharply criticized Peckham for admitting that the common law allowed reasonable restraints and then interpreting the federal antitrust statute in such a way as to make those same restraints void. Peckham, White argued, interpreted "restraint of trade" without reference to its common-law origins, where it only applied to unreasonable restraints. Thus the Sherman Act's phrase "every contract in restraint of trade" was similarly restricted when placed in its proper historical context.[52] But White's somewhat obscure critique contained a flaw: it rested on a gross distortion of the common law, particularly the nineteenth-century American common law, which had generally failed to incorporate the rule of reason into the law of non-ancillary restraints.

White was more effective when he criticized the practical difficulties in applying Peckham's inflexible literal interpretation of the Sherman Act. He noted that innumerable business contracts imply some incidental restraints. If the Court applied the Sherman Act literally, then either "all those contracts which are the very essence of trade" would be void or the Court would have to determine, without the guidance of the rule of reason, which contracts to allow, thus subjecting the people to the "mere caprice of judicial authority."[53] White was correct in asserting that the Court could not realistically overturn "every" restraint. The notoriously vague rule of reason, however, hardly seemed the ideal standard for interpreting the Sherman Act.

The *Trans-Missouri* case marked the beginning of a long campaign by Justice White to write the rule of reason into American constitutional law. The motives behind this undertaking are obscure.[54] His record on the Supreme Court—marked by wildly fluctuating positions in such key areas as substantive due process and the federal commerce power—is more notable for disregard of precedent, rampant subjectivity, and simple confusion than any doctrinaire defense of big business.[55] Yet White was doggedly consistent in his support for the rule of reason. He apparently shared the common belief that big business was essential to industrial progress. The destruction of countless small businesses, therefore, was a necessary step toward increased efficiency rather than the death of the democratic ideal.

Less than a year later, Peckham responded to White's criticism in *United States* v. *Joint-Traffic Association* (1898), a case involving an association of thirty-one railroads that had combined to fix rates and fares between Chicago and the East Coast. A railroad attorney pressed the point that the Sherman Act must be interpreted according to the rule of reason, asserting that Peckham's literal interpretation of the Sherman Act "deprives the citizens of this country of the right, never before questioned in an English or American court, of making a large class of just and reasonable contracts, often absolutely necessary to the use of property, the transaction of business and the fair compensation of industry."[56]

Justice Peckham, again speaking for a five-man majority, rejected the defense argument and declared the Joint-Traffic Association illegal. But in doing so he refined the inflexible literal interpretation of the Sherman Act that he had advanced in the *Trans-Missouri* case.

The statute applies only to those contracts whose *direct and immediate* effect is a restraint on interstate commerce. . . . An agreement entered into for the purpose of promoting a legitimate business of an individual or corporation, with no purpose to thereby affect or restrain interstate commerce, and which does not directly restrain such commerce, is not as we think, covered by the act, although the agreement may indirectly and remotely affect that commerce.[57]

In the present case, the essential question was whether Congress could prohibit a combination to fix railroad rates even though the rates subsequently established were reasonable. Peckham ruled the restraint was direct and substantial; hence the congressional power over interstate commerce was adequate to dissolve that restraint. White and two other justices dissented without opinion.

Peckham had fashioned a workable rule for interpreting the Sherman Act. Of course the direct-indirect test involved some judicial discretion. But as White and the railroad lawyers had repeatedly pointed out, a measure of discretion was essential because the

Sherman Act's blanket prohibitions against every restraint would bring industry and commerce to a halt. Two other 1898 antitrust cases demonstrated the serviceability and essential moderation of Peckham's new standard.

In *Hopkins et al.* v. *United States* (1898), Justice Peckham, for a seven-one majority, upheld the validity of the Kansas City Livestock Exchange, an association of over three hundred individuals, who purchased livestock on a commission basis. The organization fixed minimum rates for the commissions charged by its members, but it did not engage in business itself. Although the exchange operated in both Kansas and Missouri, Peckham ruled that the association was essentially local in character. Therefore the effect of any restraints on interstate commerce imposed by the exchange were "remote and indirect."[58]

Similarly, Peckham, for the same seven-one majority, ruled that the Traders' Live Stock Exchange of Kansas City was valid in *Anderson et al.* v. *United States* (1899). This exchange consisted of individuals who purchased cattle for themselves rather than on a commission basis for others. Although the bylaws of the organization forbade dealing with nonmembers, Peckham declared the association valid because "the effect of its formation and enforcement upon interstate trade or commerce is in any event but indirect and incidental and not its purpose or object."[59]

Meanwhile, in early 1898 Judge William Howard Taft of the Sixth Circuit Court of Appeals handed down an important ruling in *United States* v. *Addystone Pipe & Steel Company,* a case involving an association of six cast-iron pipe manufacturers that controlled nearly one-third of the nation's production capacity. The combination was a flagrant example of an old-fashioned pool for dividing territories and sharing profits.[60] Nevertheless defense attorneys argued that the combination was reasonable and therefore valid. Judge Taft found the association in violation of the Sherman Act. But the case is important primarily because Taft based his decision on an extensive analysis of the common law, which neither Peckham nor White had done. They merely relied on the assertions of the railroad attorneys. Taft followed Peckham and White in noting that the common law had originally prohibited all restraint on trade. In the eighteenth century it became more flexible, allowing reasonable ancillary restraints—that is, restraints supportive of other central transactions. But Taft differed from both justices in declaring that all non-ancillary restraints had always been, and still were, void at common law: "No conventional restraint of trade can be enforced unless the covenant embodying it is merely ancillary to the main purpose of a lawful contract."[61]

Taft oversimplified the trend of the common law, which was complex and frequently contradictory, but he correctly identified the predominant tendency of the American precedents. Courts in the United States generally had not gone nearly as far as their British counterparts in relaxing prohibitions against restraints on trade. Moreover, Taft's analysis was by far the most scholarly and thorough account in the American reports at that time. Most important, it produced an interpretation of the common law leading to results similar to those produced by Peckham's direct-indirect test. Both interpretations allowed ancillary restraints that were reasonably related to the main provisions of a contract. Peckham's rule was slightly broader because it allowed non-ancillary restraints, provided that they indirectly affected interstate commerce. But the direct-indirect test was much closer to Taft's position than to the nebulous rule of reason. Both Peckham and Taft had devised reasonably objective standards to guide jurists in applying the Sherman Act.

Taft's opinion gained even more importance when the Supreme Court sustained his ruling in 1899. Justice Peckham, speaking for the first unanimous Court in a Sherman Act case, raised no objection to Taft's treatment of the common law. The defense attorneys, however, did not raise the question of reasonableness; instead, they asserted that the federal commerce power could not reach the pipe manufacturing combination. Peckham therefore based his opinion on the adequacy of the commerce power, and Justice White was able to support Peckham's decision without changing his position on the rule of reason.[62]

White, however, appeared to back away from the rule of reason in the very important *Northern Securities* case in 1904, in which the Court dissolved a holding company formed solely to consolidate the Northern Pacific and Great Northern railroads. Both White and Holmes wrote dissenting opinions, primarily arguing that the Sherman Act did not apply to "mere" stock transactions. Holmes expressed support for the *Trans-Missouri* and *Joint-Traffic* decisions in his dissent, declaring, "I accept those decisions absolutely, not only as binding upon me, but as decisions which I have no desire to criticize or abridge."[63] And White, completely reversing his position on those two cases, fully concurred in Holmes's opinion.

Ironically, Justice David J. Brewer, who had rejected the rule of reason in the *Trans-Missouri* and *Joint-Traffic* cases, now embraced that doctrine. He voted to dissolve the Northern Securities Company because it was an "unreasonable" restraint of trade, arguing that the purpose of the Sherman Act "was to place a statutory prohibition with the prescribed penalties and remedies upon those contracts which were in direct restraint of trade, unreasonable and against public policy."[64] Even with Brewer's sudden reversal, the language of the opinions in the *Northern Securities* case appeared to indicate that as of 1904, eight members of the Supreme Court had rejected the rule of reason.

During the Roosevelt administration, several attempts were made to legislate the rule of reason into federal antitrust policy. In January 1904 Senator Joseph B. Foraker, who had been so instrumental in Standard Oil's evasion of the Ohio antitrust law, introduced a measure amending the Sherman Act to permit reasonable restraints on trade, but the Senate was no more willing than the Supreme Court to adopt the rule. The interstate commerce committee failed to report the bill. Undeterred, Foraker presented a similar measure to the Senate in March 1908; it also quickly died.[65]

A bill introduced in the House of Representatives by William P. Hepburn on March 23, 1908, presented Congress with the prospect of administrative application of the rule of reason. This measure provided that corporations other than common carriers could register with the Bureau of Corporations and then submit all proposed contracts and combinations to the bureau for a determination of reasonableness. If the bureau did not declare a contract or combination unreasonable within thirty days, the government lost its right to employ the Sherman Act. The bill also provided that the government could not take action on a past unreasonable restraint of trade unless it filed suit within one year after the passage of the law. Common carriers could register with the Interstate Commerce Commission and qualify for similar treatment.[66]

Commissioner of corporations Herbert Knox Smith, who would have received the power to determine which contracts or combinations were reasonable under that bill, testified on its behalf before a subcommittee of the House Judiciary Committee. But under close questioning, Smith was unable to explain to the congressmen exactly what criteria he would use to determine reasonableness. He finally retreated to the common-law definition which offered little certainty. Other defenders of the bill, such as economist Jeremiah Jencks and National Civic Federation president Seth Low, were also unable to present to the committee any objective standard for judging contracts in restraint of trade. Under these circumstances the legislation died.[67]

On April 1, 1908, William Warner of Missouri introduced the same bill in the Senate. The measure was referred to the Judiciary Committee, but the Senate hearings also went badly for proponents of the legislation. The commitee reported negatively on the bill in January 1909, declaring, "To amend the antitrust act as suggested by this bill, would be to entirely emasculate it, and for all practical purposes render it nugatory as a remedial statute." Consequently the Senate postponed the measure indefinitely.[68]

While Congress objected to administrative application of the rule of reason, President Roosevelt stood firmly against any revision of the Sherman Act that would allow the Supreme Court to apply the test of reasonableness. In April 1908, Roosevelt informed Seth Low that he would be against "a measure inserting the word 'reason'" into the Sherman Act.[69] Two years later President Taft also strongly objected to the rule of reason. In January 1910 he told a special session of Congress that to adopt the concept of reasonable restraints "is to thrust upon the court a burden that they have no precedents to enable them to carry, and to give them a power approaching the arbitrary, the abuse of which might involve our whole judicial system in disaster."[70]

Although the Court, the Congress, and two presidents repeatedly had rejected the rule of reason, White inserted the dictum into his opinion in the *Standard Oil* case. The Chief Justice was undisturbed by the fact that the Standard Oil lawyers had not mentioned the rule at all and that consideration of the issue was superfluous to the case.

White based his argument for the rule of reason on the erroneous proposition, first advanced in the *Trans-Missouri* case, that the rule long had been an integral part of the common law of both Britain and the United States. He began by tracing the origins of the rule to the early concept of contracts in restraint of trade. He then correctly noted that by the end of the nineteenth century, the British courts had broadened the concept of reasonableness to cover non-ancillary restrictions. But the Chief Justice failed to note that when nineteenth-century American courts applied the restraint of trade concept to non-ancillary contracts, they usually refused to apply the test of reason. Unable to cite precedents to support his view of the American common law on restraint of trade, the Chief Justice resorted to vague generalization and ponderous language in a futile attempt to establish the point.[71]

White then tried to refute the government's interpretation of the Sherman Act, which was based on Peckham's opinions. The government, White declared, took the position that the federal antitrust law "leaves no room for the exercise of judgment, but simply imposes the plain duty of applying its provisions to every case within its literal language."[72] Such a literal interpretation was impossible, he argued, because the broad language of the statute declared every contract illegal that restrained trade, and all contracts imposed some restrictions. Hence judges must determine which contracts fell within the provisions of the Sherman Act. The government had not advocated the total abolition of any form of judicial discretion in antitrust cases, it had merely requested that the Court follow a long list of precedents and hold illegal contracts that substantially restrained interstate trade.[73] The *Hopkins* and *Anderson* cases clearly proved that this test permitted the court to avoid the extremes suggested by White.

After distorting the government's view of the Sherman Act, White unsuccessfully attempted to deal with the *Trans-Missouri* and *Joint-Traffic* precedents. Instead of openly overturning these decisions, he attempted to demonstrate that they were compatible with the present decision. He intentionally confused the general and technical meanings of the term *reason* to arrive at the following conclusion:

As the cases cannot, by any possible conception, be treated as authoritative without the certitude that *reason* was resorted to for the purpose of deciding them, it follows as a matter of course that it must have been held by the *light of reason*, since the conclusion could not have been otherwise reached, that the assailed contracts or agreements were within the general enumeration of the statute, and that their operations and effect brought about the restraint of trade which the statute prohibited.[74]

Then White declared that there was no actual difference between the direct-indirect test and the rule of reason. If the criterion used in a case "is the direct or indirect effect of the acts involved, then of course the rule of reason becomes the guide."[75]

White's substitution of the rule of reason for Peckham's direct-indirect test did not revolutionize the interpretation of the Sherman Act; both tests required judges to exercise some discretion in evaluating the facts of each case. Nevertheless, the rule of reason produced important changes. First, the range of judicial discretion was substantially broadened. Under Peckham's test, judges determined only whether a restraint was direct and substantial. Under White's formula, a contract or combination could impose a direct and substantial restraint and still be reasonable—that is, not detrimental to the interests of the contracting parties or the public. And judges were to determine according to their own prejudices the impact of restraints on the various interests. Second, Peckham's test restricted the size of combinations; White's did not. A combination controlling 75 percent of the interstate commerce in a given article would be a direct and substantial restraint. Yet such a combination might be judged reasonable under the rule of reason if in the Court's opinion it charged reasonable prices, produced quality goods, and observed ethical business practices.

Justice John Marshall Harlan skillfully exposed some of the problems inherent in the rule of reason in a separate concurring opinion delivered with "frequent gestures and decided animation." The opinion was sharpened by Harlan's strained personal relations with White.[76] But the Associate Justice's profound antimonopoly convictions were a far more important source of wrath than personal pique. The language of Harlan's opinion, depicting the national mood in 1890, reveals the depth of his antimonopoly sentiment.

The nation had been rid of human slavery—fortunately, as all now feel—but the conviction was universal that the country was in real danger from another kind of slavery sought to be fastened on the American people; namely, the slavery that would result from aggregations of capital in the hands of a few individuals and corporations controlling, for their own profit and advantage exclusively the entire business of the country, including the production and sale of the necessities of life.[77]

Harlan's attack on White centered on the Chief Justice's abuse of sound legal principles. Harlan noted that White had overturned a long stream of precedents, beginning with the *Trans-Missouri* and *Joint-Traffic* cases, in which the Court had forcefully and repeatedly rejected the rule of reason. In his dissenting opinion in the *Northern Securities* case, White himself had admitted the leading precedents declared that the Sherman Act prohibited all interstate contracts in restraint of trade "whether they were reasonable or unreasonable."[78] Yet the Chief Justice now claimed that reasonable restraints were valid under the Sherman Act. What is more, White refused to admit that his new position contradicted the earlier decisions.

Harlan also asserted that White had replaced a workable set of precedents with the intolerably vague standard of reasonableness, a test that the Court would have to redefine in each case. Harlan quoted with approval the Senate Judiciary Committee's characterization of that standard in 1909: "The injection of the rule of reasonableness or unreasonableness would lead to the greatest variableness and uncertainty in the enforcement of the law. The defense of reasonable restraint would be made in every case, and there would be as many different rules of reasonableness as cases, courts and juries."[79] The resulting uncertainty, Harlan predicted, would throw the business community into confusion and prompt widespread harassing litigation.

Finally, Harlan rebuked the Chief Justice for engaging in judicial legislation by amending the antitrust act in a manner that Congress had repeatedly rejected. White's action amounted to "the usurpation by the judicial branch of the government of the functions of the legislative department."[80] Harlan declared that this last tendency had become distressingly common in recent years and posed a fundamental threat to the traditional concept of separation of powers. "I feel bound to say," Harlan added ominously, "that what the court has said may well cause some alarm for the integrity of our institutions."[81]

Harlan demolished White's case for the rule of reason, yet the Chief Justice won the support of seven other members of the Court. Economic considerations apparently convinced Holmes and Hughes that the rule of reason was necessary. Holmes, who viewed the great industrial combinations as manifestations of economic progress, considered the Sherman Act "humbug based on economic ignorance and incompetence." To him, White's thought was "profound, especially in the legislative direction."[82] Perhaps Holmes viewed the rule of reason as a means of reconciling the Sherman Act with the economic forces toward combination that he considered inevitable. Hughes expressed his belief in the necessity of the rule of reason several years after the Standard Oil case: "It is manifest that if the Anti-Trust Act had received a literal interpretation and had been regarded as condemning all contracts which might produce any restraint of interstate trade, it would have hopelessly tied up our commercial activities, and most appropriate business relations would have become impossible if an act so interpreted could have been upheld as constitutional."[83]

White, Holmes, and Hughes were the most influential members of the Supreme Court in 1911. Once he had enlisted the support of Holmes and Hughes, Chief Justice White was in a good position to bring the other members of the Court over to his side.[84] Furthermore, three recently appointed Justices—Lurton, Van Devanter, and Lamar—had not participated in any of the earlier Supreme Court decisions involving the rule. Apart from White and Holmes, only Justices Day and McKenna had to reverse themselves on this issue.

Justice Harlan's personality also contributed to his failure to attract support. Harlan, upset because he had been denied the coveted post of Chief Justice, was at odds with both White and Holmes in the spring of 1911. As Charles Evans Hughes put it:

Justice Harlan and Justice White did not like each other. Justice Harlan was antipathetic to Justice Holmes, and Holmes to Harlan, though each respected the soldierly qualities of the other. When in conference Justice Harlan would express himself rather sharply in answer to what Justice Holmes would say, the latter, always urbane, would refer to Justice Harlan as "my lion-hearted friend."[85]

Hughes himself considered Harlan overly combative, condemning Harlan's oral statement in the Standard Oil case as an "outburst seldom if ever equal in the annals of the Court, . . . [Harlan] went far beyond his written opinion, launching out into bitter invective, which I thought most unseemly."[86] In 1911 Harlan was seventy-eight; perhaps some members of the Court hesitated to side with the aging lion against the new Chief Justice and his highly respected allies.

Two months later, the Court demonstrated the potential of the newly proclaimed rule of reason. In *United States* v. *American Tobacco Company,* Chief Justice White declared, for a unanimous court, that the tobacco trust, which controlled over sixty subsidiaries, had violated sections 1 and 2 of the Sherman Act, but he declined to order the unqualified dissolution of the combination because a literal application of the law would violate property rights and fail to restore reasonable competition to the tobacco industry. As White saw it, reason was "plainly required in order to give effect to the remedial purposes which the act under consideration contemplates, and to prevent that act from destroying all liberty of contract and all substantial right of trade."[87] In this case, White deduced that reason dictated that the lower court devise "some plan or method of dissolving the combination and of recreating, out of the elements now composing it, a new condition which shall be honestly in harmony with and not repugnant to the law."[88]

Harlan concurred in finding the American Tobacco Company in violation of the Sherman Act. In a separate opinion, however, he again denounced the rule of reason as an inexcusable act of judicial legislation. He also objected to White's plan for restoring "reasonable" competition to the tobacco industry. "I confess my inability to find, in the history of this combination, anything to justify the wish that a new condition should be 're-created' out of the mischievous elements that compose the present combination."[89] Harlan thought that the court should have formulated "specific directions" to dissolve the trust.

Subsequent events validated Harlan's skepticism over White's scheme to restore reasonable competition to the tobacco industry. The circuit court allowed the tobacco trust to submit a dissolution proposal. The Justice Department then suggested minor modifications, and the circuit court approved the plan in November 1911. Although the decree created fourteen successor companies, the industry remained highly concentrated because certain companies retained overwhelming dominance in certain areas of the industry. For example, the American Tobacco Company dominated the most expensive and the cheapest grades of Turkish cigarettes, the P. Lorillard Company controlled the middle Turkish grades, and the Ligget and Myers Tobacco Company controlled the manufacture of most domestic cigarettes. Moreover, stock ownership remained concentrated in a few hands. The twenty-five leading stockholders held between 28 and 45 percent of the voting stock in each of the successor companies.[90]

In his *Addystone Pipe* decision of 1898, Judge William Howard Taft had concluded that according to common law, the rule of reason applied only to ancillary restraints. As president, Taft recently had rejected the rule in his annual message of 1910. Hence the chief executive might have been expected to join Harlan in condemning White's *Standard Oil* and *American Tobacco* decisions. Instead he supported the Chief Justice, declaring that the Court had "laid down a line of distinction which is not difficult for honest and intelligent business men to follow." He even told Congress that the *Standard Oil* and *American Tobacco* decisions "do not depart in any substantial way from previous decisions."[91]

Several reasons lay behind Taft's sudden reversal. He was already displaying the dislike of dissent that became so prominent in his career as Chief Justice. The day following the Standard Oil decision, Taft remarked that Harlan had delivered a "nasty, carping and demagogic opinion, directed at the Chief Justice and intended to furnish LaFollette and his crowd with as much pabulum as possible."[92] The president also favored industrial combination and therefore had no objection to the rule of reason's potential benefits to big business. Most important, Taft had to support the *Standard Oil* and *American Tobacco* decisions for political reasons. He had continued Roosevelt's policy of cultivating public hostility toward Standard Oil, and the two cases were the center-pieces of his trust-busting campaign. If Taft had objected to these decisions, he would have minimized one of the major achievements of his unspectacular term of office.

Wickersham's reaction to the rule of reason was also determined by the necessity of protecting the administration's political position and his own reputation. In his initial comment on the decision, Wickersham declared: "Substantially every position contended for by the Government in this case is affirmed by the Supreme Court." Wickersham praised the rule of reason in July 1911 before the Michigan State Bar Association: "The area of uncertainty in the law has been greatly narrowed and its scope and effect have been pretty clearly defined." The attorney general had no interest in exploring the potentially damaging aspects of the decision. When Albert H. Walker, an authority on antitrust law, sent in a critical analysis of the rule, Wickersham merely thanked him and noted that he had "not yet had time to read it."[93]

Apart from the administration and its close supporters, few leading politicians were willing to support Chief Justice White's position. Writing in *The Outlook*, Theodore Roosevelt declared: "As construed by the Supreme Court . . . the Anti-Trust Law is radically and vitally defective, and any effort to strengthen it would be worse that futile."[94] Senator Robert C. La Follette of Wisconsin, who was second only to Roosevelt in the progressive wing of the GOP, also condemned the rule of reason. Present in the Supreme Court chamber when White delivered his opinion, La Follette told reporters, "I fear that the court has done what the trusts wanted it to do, and what Congress has steadily refused to do."[95] He immediately began work on a bill that attempted to define "reasonable" restraints and thus limit judicial discretion. His measure placed the burden of proof to establish the reason-ableness of a restraint on the accused and declared that unfair competitive practices—such as exclusive sales contracts and local price discrimination—were by definition unreasonable. It also defined a 40 percent share of an industry as an unreasonable restraint. La Follette presented the bill to the Senate on August 19, 1911, but it failed to attract significant support.[96]

The Democrats reacted even more negatively to the rule of reason than the Progressive Republicans did. William Jennings Bryan asserted that Chief Justice White had "waited 15 years to throw his protecting arms around the trusts and tell them how to escape."[97] Between May 17 and July 27, 1911, three Democratic senators and three Democratic congressmen introduced measures overturning or restricting the application of the rule of reason, all failed. Democratic hostility to White's dictum remained strong during the election of 1912. The party platform declared: "We regret that the Sherman anti-trust law has received a judicial application depriving it of much of its efficiency and we favor the enactment of legislation which will restore to the statute the strength of which it has been deprived by such interpretation."[98]

Woodrow Wilson, the 1912 Democratic presidential candidate, had not developed a coherent position on the trust issue when he became the party's standard-bearer. But during the course of the campaign, he hastily developed a position on this critical issue under the guidance of Louis D. Brandeis. Brandeis rejected the common notion that trusts were both efficient and inevitable, and he favored vigorous antitrust action. He therefore strongly disapproved of the rule of reason. In fact, he had helped La Follette prepare his bill limiting the application of White's dictum. On September 30, 1912, Brandeis wrote a long memorandum to Wilson detailing actions essential to controlling the trusts. Purging the rule of reason from the Sherman Act figured prominently in the discussion. The candidate's speeches soon reflected Brandeis's emphasis on the necessity of vigorous antitrust enforcement. When the voters registered their approval of the Wilson-Brandeis approach at the polls in November, the future of Chief Justice White's rule of reason appeared uncertain.[99]

Shortly before Wilson took office, Senator Albert B. Cummins orchestrated yet another attack on the rule of reason. Cummins was chairman of the Senate Interstate Commerce Committee, which had held extensive hearings on the trust problem late in 1911 and early in 1912. On February 26, 1913, he issued the committee's report, which called for supplementary legislation to restore vigor to the Sherman Act. Such legislation was imperative because the rule of reason not only distorted the common law, it granted the courts "the vast and undefined power" to determine the reasonability of restraints on trade, guided only by the social and economic prejudices of the judges.[100]

The overwhelmingly negative political reaction to the rule of reason, particularly among Democrats, suggested that the new Wilson administration and the Democratic majority in Congress would produce legislation to overturn or at least modify the unpopular dictum. In fact, Wilson found the Democratic leadership in favor of clarifying the rule when he began to consider seriously antitrust legislation in November 1913. In an address to a joint session of Congress on January 20, 1914, the president declared himself in support of "further more explicit legislative definition of the policy and meaning of the existing antitrust law."[101]

Henry D. Clayton of Alabama, chairman of the House Judiciary Committee, directed the effort to clarify the Sherman Act. Congressman Augustus Stanley of Kentucky presented a bill overturning the rule of reason to the committee, but surprisingly the measure ended up in the committee's "capacious wastebasket." Congressman Charles Carlin of Virginia, a member of the committee, explained:

We have not changed the "rule of reason" because we found upon investigation of every decision of the Supreme Court that the men who railed against reason had lost their bearings in the forest and were groping in the dark; because the court in applying the "rule of reason," has applied it in the interests of the people, and there has never been a combination sought to be dissolved up to this moment that has not been dissolved by the application of the "rule of reason."[102]

This statement is a surprising contradiction of the Democratic platform of 1912 and is impossible to reconcile with the *American Tobacco* decision.

Instead of repealing the suddenly respectable rule of reason, the House Judiciary Committee decided to clarify the Sherman Act by prohibiting certain trade practices and forms of corporate organization. On April 14, 1914, Chairman Clayton introduced a bill outlawing discriminatory price cutting, exclusive dealing contracts, tie-in sales contracts, holding companies, and interlocking directorates. The measure also made owners and directors of corporations and businesses criminally responsible for civil violations of the antitrust laws and gave private parties suing for damages under the Sherman Act the benefit of judgments in government prosecutions. On June 5, 1914, the House approved the measure by a vote of 275 to 54.[103]

By the time the House had passed the Clayton Act, Wilson had decided that a statutory definition of all conceivable restraints on trade was impossible. He now felt that a strong regulatory commission was the most effective method of combating monopoly and lost interest in the Clayton measure; instead he turned his attention to legislation that ultimately resulted in the establishment of the Federal Trade Commission. The Clayton Act, deprived of active administration support, emerged from the House-Senate Conference Committee on September 23, 1914, in a gravely weakened state. The conferees had removed criminal provisions for civil violations of the antitrust laws. Moreover, the prohibitions against unfair trade practices, holding companies, and interlocking directorates now applied only "where the effect may be to substantially lessen competition or to create a monopoly," vague terminology that would require judicial interpretation.[104] Thus the final version of the bill reintroduced the broad judicial discretion that it had been originally designed to prevent.

Senator James A. Reed of Missouri, who had introduced a bill against the rule of reason in May 1911, bitterly remarked that the Clayton Act should be retitled, "An apology to unlawful restraints and monopolies."[105] Nevertheless Congress promptly passed the bill, and the president signed it into law on October 15, 1914. Wilson and the Democrats neither repealed the rule of reason nor clarified the federal antitrust law. The Court would still have to interpret the vague language of the Sherman Act according to the illusive concept of reasonableness.

Three subsequent Supreme Court decisions demonstrated the potential of the unrestricted rule of reason. In 1918 the Court , by a four to three vote, sustained the legality of the United Shoe Machinery Company, a firm that controlled 95 percent of the shoe manufacturing machinery in the United States. Justice Joseph McKenna, for the majority, admitted that the "company indeed has magnitude, but it is at once the result and the cause of efficiency, and the charge that it has been oppressively used is not sustained."[106] McKenna also approved of tying agreements imposed on lessees of shoe machinery because he judged that these agreements were fair to both the combination's customers and its potential competitors. He seemed unconcerned that United Shoe Machinery's overwhelming dominance of the industry, in addition to the tying agreements, gave the company near absolute power to prevent competition.

Justice William R. Day, who like McKenna had concurred in Chief Justice White's Standard Oil opinion, dissented. Day agreed with McKenna that the Sherman Act condemned only unreasonable contracts and combinations. But it would require far more than the rule of reason to legitimatize United Shoe Machinery's tying agreements. As Day saw it, those agreements remained "clearly in restraint of interstate trade and tend to monopolize in the sense that those terms have been defined by this court."[107]

Justice John H. Clarke also wrote a dissenting opinion. Like Day, Clarke had no quarrel with the rule of reason. Rather he rejected McKenna's benign characterization of the shoe machinery trust. "A careful study of this record" convinced him "that the United Shoe Machinery Company is a combination in restraint of trade and commerce."[108] Citing the *Standard Oil* and *American Tobacco* decisions, Clarke declared that the Court should employ the same "resolute manner" in dealing with this trust as it had in handling previous offenders. Three different judges had looked at the same facts and had derived three separate definitions of reasonable restraints. After the *Shoe Machinery* case, no one could doubt the ability of the rule of reason to confuse.

In 1920 the Court voted four to three against the dissolution of the United States Steel Corporation. The steel combination was a union of about 180 firms that in 1911 controlled 50 percent of the nation's steel production. It was the largest industrial merger in the

United States. McKenna, again speaking for the majority, determined that the organizers of the steel combination had neither aimed at nor achieved a monopoly. Moreover, the corporation had long before abandoned all illegal practices. McKenna declared that size was no problem:

The Corporation is undoubtedly of impressive size and it takes an effort of resolution not to be affected by or to exaggerate its influence. But we must adhere to the law and the law does not make mere size an offense or the existence of unexerted power an offense . . . . It does not compel competition nor require all that is possible.[109]

McKenna made it clear that his concept of sound public policy was the controlling factor in the decision. He was unable to see any public benefit in dissolving the company. In fact, smaller, more competitive steel producers might actually increase the cost to the consumer.

Justice Day wrote the dissenting opinion. Still adhering to the rule of reason, Day admitted that the Sherman Act did not limit the "mere size" of a corporation. Nevertheless, that fact did not justify McKenna's total disregard of the implications of United States Steel's size on competition:

From the earliest decisions of this court it has been declared that it was the effective power of such organizations to control and restrain competition and the freedom of trade that Congress intended to limit and control. That the exercise of the power may be withheld, or extended with forebearing benevolence, does not place such combinations beyond the authority of the statute which was intended to prohibit their formation.[110]

Day asserted that allowing United States Steel to remain intact because the corporation fit the majority's conception of the public interest amounted to "practical nullification" of federal antitrust legislation.

Following the *United States Steel* decision, the government dropped several antitrust cases, and few new prosecutions were undertaken during the 1920s.[111] The *International Harvester* case in 1927 demonstrated the futility of antitrust action as a method of curbing corporate combination. International Harvester controlled 64 percent of the farm implement market, and a Federal Trade Commission investigation had uncovered numerous instances of the company's anticompetitive practices. But Justice Edward Sanford, who wrote the unanimous opinion, refused to consider the Federal Trade Commission evidence, declaring it the product of "an ex parte investigation." Following the dictum of the *United States Steel* case, Sanford then asserted that the law did not "make the mere size of a corporation, however impressive, or the existence of unexerted power on its part, an offense, when unaccompanied by unlawful conduct in the exercise of its power."[112]

The "practical nullification" of the Sherman Act through the rule of reason contributed to the legal environment that permitted a new wave of corporate combination during the 1920s. Between 1921 and 1928, nearly forty-eight hundred businesses disappeared through merger or direct acquisition in the mining and manufacturing industries. Over a thousand mergers took place during those years. The two hundred largest manufacturing corporations increased their share of total corporate assets from 33 percent in 1909, to 48 percent in 1929, and to 55 percent in 1933. These same firms increased their share of total corporate income from 33 percent in 1920 to 43 percent in 1929. The top 5 percent nonfinancial corporations in size received 79 percent of total nonfinancial corporate income in 1918 and 88 percent of that same income in 1932.[113]

The rule of reason also had a symbolic importance. In 1911 the Supreme Court officially repudiated the idea that the independent small businessman was both inherently valuable to the political health of a democracy and necessary to an economy theoretically based on competition. This concept was central to the philosophy behind the Sherman Act, and it was very important to older jurists like Peckham and Harlan, but it was out of touch with economic reality. By 1911 the corporate giants clearly dominated the American economy. The rule of reason was the official constitutional acknowledgment of that fact.

(7)
Dissolution

Political reaction in Washington to Chief Justice White's Standard Oil opinion focused on the rule of reason, but the oil trust directors in New York fixed their attention on the dissolution decree. Over the years, the Standard leaders and their attorneys had perfected several methods of working around judicial directives. Corporate reorganization, political and economic pressure, and delay all had proven highly effective. In 1911, however, the oil barons made no attempt to evade either the letter or the spirit of the decree. Within a few days after the Supreme Court decison, the Standard Oil directors decided to cooperate fully with the government in restructuring the petroleum industry.

The startling weakness of the dissolution decree was the overriding reason for this cooperative attitude. The decree ordered Jersey Standard stock divided into separate blocks for each of the successor firms, but a handful of individuals still would hold a majority interest in all the newly independent companies. The government proposal created nonintegrated firms that would have to rely on other members of the group for vital services. Moreover, the plan did not disturb the old system of marketing territories. Mortimer F. Elliott, head of the Jersey Standard legal department, privately assessed the affect of the decree on the oil trust's operations:

We will be able to continue business as before, except that officers of the different companies will control the business, instead of the Standard Oil

Company of New Jersey. As I have stated for publication two or three days after the decision the different companies will continue to do the same business as heretofore, under the management of their officers.[1]

The Standard directors considered the combination's poor public image as they decided on their response to the Supreme Court decree. The oil trust had faced a hostile public ever since Henry D. Lloyd published "The Story of a Great Monopoly" in the *Atlantic Monthly* in 1881. And although the combination had manipulated the press in Ohio and Kansas through the Jennings Advertising Agency, it had never launched a national public relations effort. About the time of the Bureau of Corporations' first highly critical report on the combination, the Standard leadership finally decided that a systematic public relations program was required. In May 1906 the Jersey Standard directors appointed J. I. C. Clark to its legal staff and placed him in charge of the "newspaper department." Clark was an experienced journalist, who had served for several years as an editor for the *New York Tribune*. A former playwright and a drama critic, he had numerous contacts among opinion molders and knew how to present effectively Standard's version of the news. He reportedly received "an enormous salary" for meeting regularly with reporters, compiling newspaper releases, and artfully dodging embarrassing topics.[2]

At the same time, formerly secretive corporate officials took a more active role in creating a favorable public image of Standard Oil. Henry H. Rogers presented the company's point of view to a *New York Times* reporter in 1906. John D. Archbold issued a widely circulated statement, "To the Press and Public," in September 1907, warning the public about inaccurate negative information concerning the combination. Archbold also wrote an article for the December 7, 1907, issue of the *Saturday Evening Post*. Lesser company officials increasingly granted interviews, wrote articles, and issued denials of false or misleading stories about the organization. In 1906 a special executive committee of Jersey Standard commissioned the Reverend Leonard Woolsey Bacon to write a history of the combination. Several older Standard officials, including John D. Rockefeller and Henry M. Flagler, actively participated in the project. Bacon became ill in 1907, and a pamphlet covering the *History of the South Improvement Company* was the only portion of the work ever to appear in print. But Rockefeller was so concerned about public hostility toward Standard Oil that he changed his long-standing insistence on privacy and presented his version of the early years of the oil industry in "Random Reminiscences of Men and Events." This work first appeared in serial form in *The World's Work* beginning in October 1908. It was published as a book in 1909.[3]

This ongoing effort to defuse public hostility directly influenced the Standard leadership's response to the decree. The oil barons were anxious to halt the seemingly endless procession of critical reports in the press about their organization. Formal dissolution would replace the highly visible Jersey Standard holding company with an informal working coalition among the various former subsidiaries. The magnates hoped this arrangement would lessen public interest in the Standard group. At the very least, compliance would end the current wave of government investigations and antitrust suits, and that alone would contribute substantially to an improved image.

The Standard directors also considered the competitive situation in the oil business. As the figures in table 2 indicate, rapid expansion and radical geographic change were the dominant characteristics of oil production in the first decade of the twentieth century. In 1900 the Appalachian and Ohio-Indiana fields produced 93 percent of the nation's crude oil. By 1911 California produced 36 percent, mid-continent 30 percent, Illinois 14 percent, and the Gulf Coast 5 percent of the oil in the United States, while the older Appalachian and Ohio-Indiana fields together accounted for only 13 percent of the national total.[4] Based in these older fields, the Standard organization could not match the rate of growth set by the industry as a whole. Standard Oil remained dominant in 1911. But as table 3 demonstrates, its position had been significantly undermined.

The discovery of flush fields, yielding large amounts of crude oil in short periods of time, also stimulated competition and weakened Standard's stranglehold on the industry. The legal basis of this phenomenon was the rule of capture: surface landowners held all subsurface mineral rights. A major oil discovery, therefore,

TABLE 2
U.S. PETROLEUM PRODUCTION, 1900–1911
(IN MILLIONS OF 42-GALLON BARRELS)

Year	Appalachian	Ohio-Indiana	Illinois	Mid-Continent	Gulf	California	Other	Total
1901	33.6	21.9		0.8	3.7	8.8	0.5	69.3
1903	31.6	24.1		1.6	18.4	24.4	0.5	100.6
1905	29.4	22.3	0.2	12.5	36.5	33.4	0.4	134.7
1907	25.3	13.1	24.3	46.8	16.4	39.7	0.3	165.9
1909	26.5	8.2	30.9	50.8	10.9	55.5	0.3	183.1
1911	23.7	6.2	31.3	66.6	11.0	81.1	0.4	220.3

SOURCE: Harold F. Williamson et al., *The American Petroleum Industry: The Age of Energy* (Evanston: Northwestern University Press, 1963), p. 16.

prompted every landowner in the strike area to drill wells in an attempt to beat their neighbors to the oil reserves. Uncontrolled production and wasteful exploitation of oil resources usually resulted, but the massive flood of crude provided opportunities for new firms to enter the industry. Gulf Oil, the Texas Company, and Sun Oil all originated as a result of the flush fields on the Gulf Coast between 1901 and 1906.[5]

The different quality of the crude from the new fields also presented independent oil companies with opportunities. Standard Oil had traditionally emphasized the production of highly profitable illuminating oils, and eastern crude yielded a large percentage of these. But crude from the Gulf Coast and California was far less suitable for this purpose. That fact contributed to the Standard decision not to make a major effort to dominate the Gulf Coast field. It also explains why the oil trust was content to share the California field with Union Oil and the Associated Oil Company.[6]

The hitherto unfamiliar properties of Gulf Coast and California crude prompted the development of new uses of petroleum that in turn provided additional opportunities for independents. The heavy crude oil from the new fields made excellent fuel with little or no refining. As early as 1901, Gulf Coast producers exported substantial quantities of crude to Great Britain for use as fuel. Gulf

TABLE 3
SUMMARY OF STANDARD OIL'S POSITION
IN THE AMERICAN PETROLEUM INDUSTRY
1880–1911

PERCENTAGE CONTROL OVER CRUDE OIL SUPPLIES

FIELDS	1880	1899	1906	1911
Appalachian	92	88	72	78
Ohio-Indiana		85	95	90
Gulf Coast			10	10
Mid-continent			45	44
Illinois			100	83
California			29	29

PERCENTAGE CONTROL OVER REFINERY CAPACITY

	1880	1899	1906	1911
Share of rated daily crude capacity	90–95	82	70	64

PERCENTAGE OF MAJOR PRODUCTS SOLD

	1880	1899	1906–1911
Kerosene	90–95	85	75
Lubes		40	55
Waxes		50	67
Fuel oil		85	31
Gasoline		85	66

SOURCE: Ralph Andreano, "The Emergence of New Competition in the American Petroleum Industry Before 1911" (Ph.D. diss., Northwestern University, 1960), p. 282.

Coast sugar refineries, brick works, and breweries quickly converted from coal to fuel oil. Railroads in the Southwest and on the Pacific Coast rapidly became the most important consumer of fuel oil. And the United States Navy's Pacific fleet used 750,000 barrels of California oil as fuel in 1911. By that time fully 57 percent of national crude production was being used as fuel oil. Independent producers in California also developed a substantial market for asphalt, which was widely used in industry and as road paving and roofing.[7]

At the time of Chief Justice White's decision, Standard Oil remained a rapidly growing and immensely successful concern. As the dominant force in a dynamic industry, it retained the power to manipulate prices, extract vast profits, and crush many of its competitors. But its position was eroding, and the Standard leadership appreciated that fact. A trifle more independence for subsidiaries under the Supreme Court directive might actually promote managerial initiative and help stem the rising tide of competition without damaging the essential unity of the combination.

Like the oil barons, Kellogg and Wickersham were pleased with the action of the Supreme Court. Neither their private correspondence at the time of the decision nor their later public statements contain any hint of dissatisfaction over the weakness of the dissolution proposal. In 1912 Kellogg strongly endorsed the high court's directive in the *Review of Reviews:* "In my opinion that decree accomplished everything that is possible to accomplish under the Sherman Act." About the same time Attorney General Wickersham announced that the decree reached every conceivable goal "short of absolute confiscation."[8]

Blessed with an overabundance of good feelings on both sides, the trustbusters and the Standard lawyers implemented the Supreme Court directive with little difficulty, and the Justice Department was more than accommodating in ironing out the few minor problems that surfaced. On June 9, 1911, Standard attorney John Milburn wrote to the attorney general about the possibility of an extension after the six-month time limit for dissolution had expired: "We fully expect to get through within the six months allowed, which was the minimum time fixed by the Supreme Court.

However doing our best, we may find ourselves in the position of requiring a little more time."[9] A few days later, Milburn asked for assurance that the langugage of the decree guaranteed that "during the six month period whilst the arrangements for the distribution of stocks are being perfected the present conditions must necessarily continue." Through this request he wanted to make certain that the Jersey Standard directors would have the power to vote the stock of the various subsidiaries and to adjust the price of that stock to represent better the real value of the various companies.[10]

Kellogg suspected Milburn's motives and advised the attorney general against granting either request. He was particularly concerned about the oil barons' voting the stock of the subsidiaries during the six-month dissolution process: "I cannot see why, during that time, in addition to increasing or decreasing stocks, they may not go ahead to consolidate a lot of these corporations."[11] But the attorney general decided to leave open the possibility of stock value adjustment and refused to rule out a time extension: "On the question of time, he [Milburn] called my attention to the fact that the Supreme Court said they should have at least six months time." In a letter to Kellogg explaining his cooperative attitude, Wickersham spoke glowingly of the "good faith" of the Standard attorneys and expressed his belief that the oil trust would be dismantled without significant delay.[12]

Kellogg filed the decree in the federal circuit court at St. Louis on June 21, 1911, and the six-month dissolution period began on that date. Wickersham's last-minute concessions to Milburn did not significantly weaken the already fatally defective decree. The Jersey Standard board of directors announced their plan for a pro-rata distribution of stock on July 28, and they completed the distribution for domestic affiliates within the six-month limit. Standard officials, however, required and received a small extension to complete the distribution of Anglo-American stock, which they did on January 20, 1912. Standard officials made no effort to adjust stock values during the stock distribution process.[13]

Because Attorney General Wickersham had faith in both the efficacy of the dissolution decree and the good intentions of the Standard leaders, he opposed criminal action against the oil barons just as firmly as earlier federal officials had. On November 29, 1909, Herbert S. Hadley, who had conducted the Missouri suit against

Standard Oil, wrote to the attorney general recommending criminal suits: "When we consider that in the organization and the conduct of this business that those who have been responsible therefore have left a record of deceit, trickery, oppression and injustice that is a disgrace to the commercial history of this country, the necessity of criminal prosecutions, to my mind, is strongly apparent." He added that there was "substantial evidence to establish the truth of most of these charges," and he referred Wickersham to the voluminous record of the Missouri suit for that evidence.[14]

Wickersham replied that he had not considered criminal prosecutions at that point "because of the pendency of the civil suit" against Standard Oil: "Personally, I am strongly disposed to think that the criminal features of the case ought not to be taken up until after a final decision by the Supreme Court." The attorney general noted that a successful resolution of the civil suit would greatly enhance the prospects for a successful criminal prosecution, and he promised to give "mature consideration" to Hadley's suggestion.[15]

But Wickersham's actions after the Standard Oil decision make it appear doubtful that he ever seriously considered criminal suits against Standard Oil. In the days immediately following that decision, many complaints poured into the Justice Department from independent oil dealers and interested citizens urging criminal charges against the Standard leaders, but the Justice Department consistently refused to take action.[16] On May 23 the United States Senate passed a resolution demanding to know "what, if any, criminal prosecutions have been begun, or are now pending against the said Standard Oil Company of New Jersey, or the said constituent companies, or individual defendants." To this Wickersham replied: "I am directed by the President to inform you that no criminal prosecutions have been begun."[17] The administration considered the dissolution decree adequate to end Standard Oil's domination of the petroleum industry.

The Taft administration may have been satisfied by the dissolution decree, but a substantial portion of the press was skeptical. Shortly after the Standard directors announced their stock distribution plan, the *Outlook* pointed out the most obvious deficiency:

A majority of the stock of the Standard Oil Company is owned by less than a dozen men. Since the New Jersey company owns practically the entire capitalization of the thirty-three companies, when the distribution has been made, that same group of a half score men will control each of the thirty-three companies. It is hardly to be expected of human nature that they will, through the agency of the different companies which they own, enter into violent competition with themselves.[18]

That same week, the *Literary Digest* collected and reprinted considerable newspaper comment expressing similar viewpoints.[19]

On September 7, 1911, the New York *World*, which had long delighted in attacking the oil trust, revealed another serious defect in the decree. It charged that officers with no desire for competition within the Standard group staffed the boards of directors of the recently liberated affiliates. A *World* headline exclaimed: "Oil Trust Lamp Beats Aladdin's with Its Magic. Humble Clerks, Agents and Mechanics 'Called Upstairs' in No. 26 Broadway, and Lo! They Were Directors!" The *World* went on to report that the "deadly rivals of oil" lunched with each other in "sweet peace" in the headquarters dining room. Despite the sensationalism, the *World* was substantially correct in arguing that the new managers of the former Standard subsidiaries had neither the experience nor the desire to compete with their former colleagues and superiors.[20]

Several well-publicized developments in the two years following the Supreme Court decision convinced a large segment of the public that the dissolution decree was inadequate and that the oil combination remained as united and powerful as ever. The rise in the value of Standard Oil stock was spectacular. At the time of the dissolution, Standard securities were valued at $663 million. On March 8, 1912, after the destruction of the combination, these same securities were worth $885 million, a $222 million increase within ten months.[21] The rise of Standard of Indiana stock was particularly impressive. On December 31, 1911, the value of this stock was $2,521 per share, in late January it was selling for about $4,000 per share. On February 13, 1912, Indiana directors announced that as a result of an increase in capitalization from $1 million to $30 million, stockholders would receive twenty-nine new shares for each share held. Indiana Standard stock then soared to $7,000 per share.[22]

Standard stock values rose sharply because the dissolution process revealed to investors the real assets of companies that had been undercapitalized for years. The continued rapid expansion of the petroleum industry as a whole was another important factor. But the ineffectiveness of the decree also contributed to investor confidence in the Standard companies and thus pushed up prices. C.H. Pforzheimer, the leading Wall Street specialist in Standard securities, informed his customers in November 1912:

It is clear that the disintegration has not altered appreciably the natural commercial or trade relations between the various former subsidiaries of the Standard Oil of New Jersey. Indeed they will continue to transact business with each other in the same manner as they did before the dissolution. Already it is obvious that each and every company is operated as efficiently and profitably today as before the disturbed intercorporate relations—through stock control.[23]

Little wonder that Albert W. Atwood, writing in *McClure's,* labeled the Standard dissolution "The Greatest Killing on Wall Street."[24]

Another aspect of the Standard stock distribution seized the public's attention. Immediately after the dissolution, the small stockholders received fractional shares in a total of thirty-four companies. Face values of these shares were as low as seven cents. These "splinters" replaced a single share of Jersey Standard stock worth $700. Moreover, only shareholders who owned full shares in a company could vote. Thus many small stockholders sold their shares before the dramatic rise in the stock prices. Insiders, who knew that the Standard companies were undercapitalized and that the decree was toothless, quickly bought all available splinters, producing an even greater concentration of shares in the possession of the dominant stockholders. Figures for Indiana Standard illustrate the trend. On September 1, 1911, there were 6,081 stockholders; seven months later, before the explosive rise in Indiana Standard stock prices, the number had decreased to 5,074. As Albert Atwood put it, "The government has forced the operations of great concerns more and more into the 'inside' group, and made possible the distribution of tens of millions more to multimillionaires."[25]

During the first year after dissolution, Standard companies paid record dividends. Dividends for the years 1902 through 1911 had ranged between 44 and 36 percent; in the first eleven months of

1912, the newly independent firms announced dividends averaging 53 percent of the value of their stock. The companies paid over $52 million during that period.[26]

The American public did not have to speculate about the source of these increased profits. The price of kerosene, gasoline, and most other petroleum products began to rise in 1911. In June 1912 *Everybody's Magazine* published a survey of competitive conditions in gasoline marketing in Massachusetts, Connecticut, and New Jersey that revealed almost total Standard dominance and widespread price manipulation.[27] By itself this limited magazine survey indicates more about continuing public distrust of Standard Oil than about actual competitive conditions in the industry. But several subsequent government studies confirmed that the various Standard marketing firms, each operating within its own territory, effectively set retail prices for petroleum products for several years after 1911.

A St. Louis trial early in 1912 also cast doubt on the effectiveness of the dissolution. After decades of controversy, Henry Clay Pierce, president of the Waters-Pierce Oil Company, had grown tired of his association with the Standard combination, and he viewed the dissolution of the trust as an opportunity to break free. At the annual Waters-Pierce stockholders' meeting, held shortly after the dissolution, Pierce refused to count the votes cast by Standard interests, even though those votes constituted a majority. In the resulting litigation Pierce argued that Rockefeller and other leading Standard stockholders were illegally attempting to retain control of former subsidiaries. Although Rockefeller eventually settled the dispute by selling out to Pierce, the Waters-Pierce stock fight appeared to many as yet another indication that the oil trust was still operational.[28] After reviewing this episode, the *New York Herald* declared:

The result [of the dissolution] was a paper victory for the people. The real victory . . . was won by the oil company, which is run in practically the same old way, by the same old men, with profits even greater than formerly.[29]

Standard Oil's continuing activities in Texas also captured the public's attention. In early 1913 state Attorney General B. F.

Looney discovered that Standard interests had reacquired, through the Driskill Hotel receiver's sale, oil refineries and pipelines separated from the combination by the courts in 1909. Those properties were then reorganized as the Magnolia Petroleum Company; Standard board members owned 85 percent of the new company's stock. In March 1913 Looney filed suit against Jersey Standard claiming that that company had violated the Texas antitrust law by secretly holding a controlling interest in Magnolia Petroleum. At first Standard officials stoutly denied the attorney general's accusation, but in July 1913 they admitted their guilt and arranged a compromise settlement with the state: Jersey Standard paid a $500,000 fine and placed the Magnolia stock in the hands of a trustee.[30] Once again, public distrust of the oil combination proved justified.

While the press exposed the inadequacy of the dissolution process, complaints against the oil trust continued to pour into the Justice Department. The attorney general received letters about Standard's continuing illegal contracts with railroads, domination of natural gas companies, and restrictive agreements with independent oil firms. An Ohio retailer complained that Standard refused to supply him with oil. A Nebraska merchant detailed Standard's various anticompetitive practices, and a Standard Oil stockholder observed that the dissolution decree hurt the small stockholders and helped the large ones.[31] By far the most common complaint concerned the rise in the price of petroleum products. Protests came in from all over the country on this subject. As one irate citizen put it, "It would seem as though the people are paying the expenses at both ends of the suit. The Government's end by taxes and the defendant's end by the increased cost of their commodity."[32]

Ernest D. Owen, a Chicago antitrust lawyer, sent one of the most telling complaints against the dissolution process to President Taft in December 1911. Owen could not believe that the government had consciously allowed a pro-rata stock distribution in the Standard Oil case. He hoped that there had been some sort of mistake. Taft referred the letter to Wickersham, and the attorney general vigorously defended the pro-rata distribution. After receiving Wickersham's letter supporting the stock distribution, Owen responded that the American people "will obviously say that to distribute this stock among its [Standard's] own stockholders, is only changing the form of things, and that it ends the case in the fiasco they have feared."[33]

The public gradually became aware of the complaints piling up at the Justice Department. On September 29, 1912, the Washington *Times* reported: "Rapidly accumulating evidence as to the ineffectiveness of the dissolution decree against the Standard Oil Company is gathering in the hands of the Department of Justice, in spite of the assertions of Attorney General Wickersham that the decree resulted in a real reorganization of a sort to bring about competitive conditions."[34] To combat the rapidly escalating skepticism of the press, Kellogg induced C. D. Chamberlain, secretary and general counsel for the National Petroleum Association, to issue a public statement in October 1912, declaring that the government suit had restored competition to the oil industry. Wickersham was delighted when all the New York papers carried the story. He wrote to Kellogg: "I feel very much indebted to Mr. Chamberlain for the statement which is so clear, concise and to the point, and to you for your effective aid in securing it for me."[35]

But Wickersham was becoming increasingly anxious about the course of the Standard Oil dissolution. In late 1911 he had sought reassurance from Milburn that the combination was in fact being disassembled. The Standard lawyer had sent several letters to the attorney general reiterating the oil trust's good intentions and providing specific information on the progress of the stock distribution. Milburn's soothing words temporarily eased Wickersham's doubts, but by May 1912 the attorney general could no longer contain his anxiety.[36] This time his concern focused on the effect of the decree on retail petroleum prices. On May 6 he wrote: "At the moment, for some reason which I don't understand, the retail prices of some fuel-oil has advanced; but I cannot believe that thirty different companies, with separate managements, under the normal conditions of trade which will result, can either standardize or hold up the price in face of competition which the universal experience of the past shows inevitably steps in unless there is an artificial control to prevent it from working."[37] The attorney general still stubbornly clung to his faith in the completeness of the dissolution process. In addition, he appeared to be unaware that the rapidly growing demand for petroleum products exerted upward pressure on prices.

In October 1912, shortly after Chamberlain's letter appeared in the newspapers, Wickersham decided that action was necessary on the oil situation. On October 17 he appointed Charles B. Morrison special assistant in charge of investigating Standard Oil's alleged violations of the decree and named Oliver B. Pagan as Morrison's assistant.[38] The attorney general had refused to investigate the allegations against Standard Oil for a year and a half. When he finally decided to look into the charges, he appointed the aging jurist who had helped design the decree to head the investigation. Wickersham had no particular desire to uncover the inadequacies in an antitrust suit that the Taft administration valued as one of its major achievements.

Morrison dutifully pursued his investigation throughout the final few months of the Taft administration, observing that "all the persons whom I saw believe that the combination still exists and that the Standard companies are still working in harmony."[39]   On February 28, 1913, only four days before Taft left office, the investigator submitted a ninety-seven-page report to Attorney General Wickersham. Morrison had not found evidence of any formal agreement to avoid competition, but the way Standard companies conducted business guaranteed extensive cooperation. Morrison noted that the same individuals who ran the oil trust now controlled the newly independent companies. He acknowledged that the old system of marketing territories remained unchanged, and he observed that Standard pipeline companies were still operating primarily as auxiliaries to Standard refineries. The special investigator concluded: "The defendants have not kept faith with the court and have not abandoned the combination, but, on the contrary, have gone forward with it in full force on slightly different lines. Though it is the same in substance as it was before the decree."[40]

About to leave office, Wickersham had neither the time nor the desire to pursue this matter. But Woodrow Wilson's attorney general, James C. McReynolds, directed Morrison to continue his work. Wilson had campaigned against monopoly, and the Justice Department still received many complaints against the oil trust.[41]

Hence the new attorney general had little choice but to authorize Morrison to proceed with his investigation. Moreover, McReynolds already had demonstrated his personal commitment to antitrust enforcement. In 1910 he had worked for the Justice Deparment on the American Tobacco suit, and he quit in disgust the following year when Wickersham approved of a weak dissolution plan. In fact Wilson appointed McReynolds to the cabinet on the strength of his antitrust record.[42] The special prosecutor therefore continued to gather evidence from around the country, and he met with McReynolds and other Justice Department officials throughout the spring of 1913.[43]

Morrison submitted two reports to the Justice Department during McReynold's eighteen-month term as attorney general. In June 1913 he noted that the oil trust remained intact because the same stockholders owned all the Standard companies. While acknowledging that the Supreme Court allowed the present stockholding pattern through the pro-rata distribution, Morrison observed that the decree prohibited any "express or implied" arrangement to continue the combination. He therefore recommended that the Justice Department begin contempt proceedings.[44] In February 1914 Morrison repeated his analysis of the competitive situation in the petroleum industry but backed away from his earlier unqualified recommendation of an immediate contempt suit. He now proposed that the government grant Standard Oil six more months to achieve actual dissolution before taking legal action because of the "peculiar phraseology of the decree" that allowed pro-rata stock distribution.[45]

McReynolds refused to move in any direction. In July 1913 he told Milburn, "I have not yet definitely decided on what course of action to pursue in respect of these matters."[46] A few days later, the department issued instructions regarding the Standard Oil investigation to a United States attorney in Oklahoma. "As Mr. Morrison has submitted his final report, you need do nothing further towards investigating these companies until requested to do so."[47] Although Morrison continued to receive complaints and compile evidence from around the country, the investigation languished throughout the remainder of McReynolds's stewardship at Justice. Woodrow Wilson, like Roosevelt, was far more interested in campaigning against the trusts than battling them in court.

On August 26, 1914, the Chicago *Examiner* shattered the attorney general's tranquility:

Information on which the leading figures in Standard Oil might have been criminally prosecuted has been pigeonholed in the office of Attorney General McReynolds for more than fifteen months. . . . [Morrison and Pagan] formulated their data into a comprehensive statement. Then they concluded with the statement that the companies openly, notoriously and flagrantly had violated the dissolution order. . . . The investigators kept one copy and sent the original to Attorney General McReynolds at Washington. That was fifteen months ago. Nothing has happened. The Standard Oil Company, in its branches all over the country, operates as before unmolested.[48]

Morrison, who lived in Chicago, hurriedly assured the Justice Department that he had not leaked the information. The perennial special prosecutor then expressed his hope that this flap would not damage McReynolds's pending nomination to the Supreme Court. The day after the Chicago *Examiner* story, the United States Senate passed a resolution demanding to see Morrison's report.[49] McReynolds, his political future threatened, fought to keep the report secret, explaining to President Wilson's personal secretary: "It seems quite plain to me that it would be incompatible with the public interest to send to the Senate any report made to me by these assistants, or to divulge the results of the conference between us. The subject is still under active observation and consideration, and I think it ought to be treated in harmony with the long established practice as a confidential proceeding within this Department."[50] Wilson apparently agreed with his attorney general because McReynolds succeeded in keeping the report from the Senate and shortly thereafter was appointed to the Supreme Court.

The new attorney general, Thomas W. Gregory, wanted to know the details of the Standard Oil investigation. A Texan, he had a personal interest in this matter because he had worked on the 1907 case of *Texas* v. *Waters-Pierce Oil Company* that had resulted in a $1.6

million fine.[51] Thus, on October 30, 1914, Morrison submitted yet another report detailing his various investigative efforts and reports. But his enthusiasm for contempt proceedings against Standard Oil had dimmed. In his mind, the stock distribution provision in the dissolution decree seemed a great impediment to successful court action. He concluded that because of "the difficulties which any proceeding would encounter by reason of the provision in the decree permitting a pro-rata distribution of the stocks of the constituent companies, I do not now recommend starting proceedings based on the facts now at hand, I suggest that the inquiry be continued and information concerning conditions be kept up to date."[52]

The new attorney general apparently found the policy of perpetual investigation acceptable because Morrison continued his intermittent investigation throughout 1915. By mid-year Morrison had lost all enthusiasm for the project. Since he had helped to formulate the original decree, he had little to gain by exposing the inadequacy of his earlier work. Furthermore, he had repeatedly reported Standard's circumvention of the decree and no action had ever been taken. In June 1915 he wrote: "I beg to suggest that in view of the creation of the Federal Trade Commission, it may be that you no longer need my services."[53] The department, however, urged him to continue, and in November he submitted what he hoped would be his last report. It was a brief document dealing with minor complaints against Indiana Standard. He found most of the complaints unsubstantiated and again asked to be released from the task. In December the department finally agreed to suspend the investigation.[54]

But gasoline prices continued to rise, and complaints against Standard Oil continued to flow into the Justice Department. Prodded by public pressure, Congress turned its attention to the petroleum industry in early 1916. In January the House of Representatives demanded an investigation followed by legal action to curb rising gasoline prices. In April the Senate passed another resolution demanding to see all Justice Department studies concerning Standard Oil's compliance with the decree of 1911.[55] The attorney general brought Morrison out of retirement and put him in charge of the inquiry. Like his predecessor, Gregory stood firm against disclosure of the facts about Standard Oil. He wrote to the president of the Senate: "As the matter is still pending, I am of the opinion that it would be incompatible with the public interest to send to the Senate the reports that have been made to me from time to time by my assistants engaged in this work."[56]

In June 1916 Morrison submitted yet another report to the Justice Department, detailing the same problems—all companies drew their leaders from the oil trust, pipelines were not open to the public, similar stockholders guaranteed the unity among the companies—and again concluding "that close working relations exist between all of the Standard companies, each helping the others in the conduct of the business and in combating competition."[57] But this time he did not recommend any particular course of action to remedy the situation. Although he attended the Federal Trade Commission's hearings on the petroleum industry in June 1916, he took no further significant action for over a year.[58]

Morrison's five-year investigation finally ended with his report of July 5, 1917, which concluded: "All agree that competition does not exist between the companies whose stocks were held by the Standard of New Jersey." This failure to compete resulted from the pro-rata stock distribution: "The distribution of these stocks in accordance with the provision established a common ownership in the stockholders of the New Jersey Company of the stocks of all the subsidiary companies. Such common ownership enables the stockholders to control not only the New Jersey Company but also all of those other companies whose stocks are held by the New Jersey Company prior to the decree and to prevent competition among them."[59]

As one of the original authors of the decree, Morrison was reluctant to state that the decree was fatally defective from the start: "I agree with those who are of the opinion that the decree required the combination to be dissolved within the time fixed therein." Taking this position, Morrison should have concluded that Standard Oil had been in violation of the Supreme Court directive for five years, which was the position he had taken in his first report in 1913. But the years had eroded his resolve: "Fairness and candor impel me to say that there is much force in the position taken by those who hold the contrary opinion [that the decree did not require actual competition]. The question is debatable, and in view of the conditions brought on by the War, I do not think the present is a propitious time to institute contempt proceedings."[60]

Assistant Attorney General G. Carroll Todd, who had supervised the investigation for several years, appended his comments to Morrison's final report: "All agree, as stated by Mr. Morrison, that there is little or no real competition amongst the group of companies which formerly constituted the Standard Oil combination." But Todd did not share Morrison's hesitance to place the blame for this failure on the terms of the decree itself: "The failure of the various companies of the Standard Oil group to compete with each other does not constitute ground for charging a violation of the dissolution decree, since, as just stated, that condition is due to the community of stock ownership existing amongst them, which community of stock ownership in turn was authorized by the dissolution decree itself." Under normal circumstances, Todd would have approved legal action against Standard Oil: "I agree, however, with Mr. Morrison that the present is not a propitious time to institute such proceedings. I have discussed the matter with the Attorney General and he concurs."[61]

From Morrison's first report in February 1913 to his last in June 1917, his essential conclusions remained the same: the decree had not restored true competition. These reports repeatedly documented the same abuses that were widely covered in the press. Still the Justice Department postponed action. Wickersham's refusal to move is understandable. McReynolds and Gregory's reluctance to become involved in the antitrust morass contrasted sharply with Democratic rhetoric but was in line with the Wilson administration's generally lax antitrust policy.[62] The fact was that the original Standard Oil decree was so defective that the prospects for contempt proceedings were uncertain. By 1917 everyone at the Justice Department eagerly seized upon the war as a good excuse to drop the whole matter.

Standard Oil's compliance with the decree failed to enhance its standing with the American people in the two years immediately following the Supreme Court decision, but by 1914 public hostility toward the Standard group was rapidly fading. A sharp decrease in the number of periodical articles on the Standard companies indicated the shifting public mood. After years of continuous journalistic assault, neglect was the greatest gift the press could bestow on Standard Oil. The number of entries in the *Reader's Guide to Periodical Literature* provides a rough indication of this trend. Between 1905 and 1913—the era of investigation, litigation. and dissolution—entries averaged twenty per year.[63] But in the nine-year period beginning in 1914, the average dropped to 2.7 entries per year.[63]

The general character of the articles also changed. Accounts appearing before 1914 were overwhelmingly critical, but after that, there was an increasing tendency to portray the Standard companies in a more favorable light. Several articles appeared praising the abilities of the Standard leaders and portraying the combination as a great industrial pioneer. In 1916 the *Nation* called Standard Oil "the greatest product of American constructive genius outside the sphere of government." Standard Oil of New Jersey received widespread praise for placing its workers on an eight-hour day. In 1918 the *Outlook*, which had been a frequent critic of the company, declared that Jersey Standard's labor policy placed it in "high rank among the truly progressive corporations." The company was also warmly praised for improving the standard of living around the world by its distribution of illuminating oils.[64] Admittedly some of this praise originated in conservative quarters, nevertheless, widespread praise for Standard Oil from any source was a marked departure from earlier years when journals of every ideological persuasion freely attacked the combination.

Public appreciation of the role of the petroleum industry in World War I—epitomized by Lord Curzon's remark, "The Allies floated in victory on a wave of oil"—undoubtedly played an important part in improving the Standard image.[65] The generally pro-business political climate of the 1920s also helped. But the dismantling of the oil trust, long the nation's leading symbol of predatory wealth, was an indispensable first step on the road to public acceptance. In the long run, Standard's compliance with the dissolution decree was a public relations triumph.

The decree had an equally favorable effect on the Standard group's position in the petroleum industry. As in the years before

1911, its total share of the oil business slowly declined. But the explosive growth of the industry, new flush fields, and growing demand for new products were the leading factors behind this trend.[66] To a remarkable extent, the Standard group was able to preserve its unity, sustain a high rate of growth, and maintain its dominant position in the industry while decentralizing its management in accordance with the Supreme Court directive. A series of government reports released between 1916 and 1928 publicly confirmed what the Justice Department had tried to hide: the antitrust suit had not destroyed the Standard combination.

Standard's domination of petroleum pipelines long had been recognized as the single leading source of the combination's power. Yet the Standard group continued to employ its pipelines to restrict competition for years after 1911. The Federal Trade Commission drew the following conclusions from a 1916 investigation of petroleum pipelines in the mid-continent field:

(1) There is generally a large difference between the cost of pipe-line transportation and pipe-line tarrif rates, while the independent shipper cannot use railroads instead because their rates are still higher.

(2) The pipe-line companies generally require large minimum shipments, which makes it impracticable for small producers or refiners to ship crude oil by pipe-line.

(3) The price of the crude oil delivered at the refineries is to a large extent made up of the transportation charge.

(4) The cost of pipe-line construction is so great that small concerns cannot build lines from the Mid-Continent field to the large consuming and distributing markets.

(5) Lower pipe-line rates and small minimum shipments are necessary to enable small concerns to compete with large refineries affiliated with pipe-line companies.[67]

In 1917 the Federal Trade Commission proposed a plan for breaking Standard's domination of petroleum pipelines. At the same time it indirectly criticized the stock distribution provisions of the dissolution decree. The commission recommended that no organization engaged in producing or refining oil should be allowed to own a controlling share of the stock in a petroleum pipeline company engaged in interstate commerce.[68] In 1923 a Senate subcommittee reported that the Standard pipeline companies were not operating as common carriers as the dissolution decree and the Hepburn Act had intended. These supposedly independent companies remained little more than "plant facilities, bringing the oil from the producing field into the Standard Oil Company refineries."[69] In 1928, seventeen years after the dissolution, the Federal Trade Commission renewed its call for independent ownership of the Standard pipeline firms. The commission recommended that unless the managers of the Standard pipelines serving the mid-continent region immediately altered their policies to allow independents to use those lines, "there should be absolute dissociation of pipe line ownership from interests engaged in producing and refining crude petroleum to establish free and fair competition in this branch of the petroleum industry."[70]

Standard's continued domination of the nation's pipelines made it the leading buyer of crude oil. The Standard group consequently exercised a dominant influence over crude oil prices long after 1911. In 1921 the Federal Trade Commission outlined the crude oil price situation in the California oil fields:

A practice of great significance that had become established in the Pacific coast petroleum industry is that of basing contract prices for crude petroleum upon the market price as announced by the Standard Oil Company and by providing that an increase or decrease in the Standard's announced price during the life of the contract shall cause a corresponding change in the price paid under the contract. The practice of making such contracts is quite general for the purchase of crude petroleum, and also in the sale of fuel oil, and such contracts are made to some extent for the sale of other petroleum products. The significance of this practice is that a price change by the Standard automatically causes the same change in the prices of the other companies.[71]

A 1928 commission report indicated that Standard companies effectively set crude oil prices in the other producing fields well into the 1920s.[72]

Standard marketing companies maintained their system of exclusive sales territories for a considerable time after 1911. In 1917 the Federal Trade Commission reported that Standard gasoline marketers "maintained a complete division of territory" that embraced the entire country. It also noted that "almost without exception each Standard marketing company occupies and supplies a distinct and arbitrarily bounded territory."[73] Two years later the United States Fuel Administration declared that the Standard marketing companies competed "under no circumstances . . . even when a state line divides a town or community." The marketing units limited their activities to "the territory in which they were in operation at the decree in the Standard Oil case."[74] In 1923 the Senate Subcommittee on Manufactures reached similar conclusions following a lengthy investigation.[75]

The Fuel Administration also observed that the lack of competition among Standard marketing companies frequently led to substantial and unjustified variations in both wholesale and retail prices of gasoline. For example, on January 1, 1919, the wholesale price of Jersey Standard gasoline in New Jersey was $.225 per gallon; on the other side of the Hudson, New York Standard's wholesale price was $.245 per gallon. On the same date Ohio Standard's retail price for gasoline was $.255 per gallon; across the state line in Pennsylvania, Atlantic Refining retailed gasoline at $.28.[76]

There were several reasons for the lack of competition between Standard marketing companies. The Federal Trade Commission concluded that the community of interest existing among the companies based on common stockholders was the most important factor. Moreover, firms with excess refining capacity, such as Jersey Standard and Indiana Standard, frequently supplied other Standard marketers with products. These companies therefore had nothing to gain by competing at the retail level with companies that they themselves supplied. Finally, confusion over the use of registered brand names impeded competition. Before the dissolution each marketing subsidiary had sold Standard brands within its own territory, and Standard officials doubted that they could use these registered brands outside their original sales area. The dissolution decree had not created a marketing structure conducive to genuine competition.[77]

By scrupulously avoiding competition with each other, the Standard marketing companies remained strong enough to set prevailing retail prices in their respective areas. In 1917 the Federal Trade Commission declared: "The various Standard companies, with relatively unimportant exceptions, in announcing their tank-wagon price of gasoline in any locality, practically fixed the price that prevails."[78] In 1920 the commission similarly observed that "price initiative today seems to be left generally to the Standard companies."[79] Finally, the commission concluded that as late as 1927 Standard marketing firms still served as price leaders: "The consensus of opinion in the trade . . . is that the Standard Oil Companies establish the tank-wagon and filling station prices of gasoline, which other marketers follow as a general rule."[80]

Indiana Standard's handling of the Burton patent for thermal pressure cracking is a final example of the continuing cohesion of the Standard group after 1911. In 1909 William M. Burton, an Indiana Standard research chemist, began experiments in refining techniques aimed at substantially increasing the gasoline yield of crude oil. Supported by a company investment of $800,000, Burton and his associates perfected a method of heating crude oil under pressure in order to increase gasoline yields in 1913. He patented his process and assigned the rights to Indiana Standard. Late in 1913 the company directors decided to license the Burton method to other Standard companies but to deny outsiders access to the process. Indiana Standard required the licensees to restrict their sales of cracked gasoline to their respective marketing territories. In 1921 Indiana Standard lost its exclusive rights to the process through litigation, but Burton's invention had given the Standard group an important advantage over its competitors for almost a decade.[81]

In 1917 the Federal Trade Commission commented on the justifiability of its use of the term "Standard Oil" to discuss the various disaffiliated companies of the former trust: "An examination of the lists of stockholders of the various companies called 'Standard' shows that they are owned by bodies of stockholders which are so similar in membership as to justify the common usage."[82] Six years later the report of the Senate Subcommittee on Manufactures proclaimed: "The dominating fact in the oil industry today is its complete control by the Standard companies."[83]

(8)
Trust Busting and Public Relations

In the closing pages of *Wealth Against Commonwealth*, an impassioned Henry Demarest Lloyd wrote: "Democracy is not a lie. There live in the body of commonalty the unexhausted virtue and the ever-refreshed strength which can rise equal to any problems of progress."[1] In the late 1880s many Americans, including Lloyd, entertained the belief that antitrust legislation might be an effective expression of that democratic strength and virtue. The Standard Oil cases were a significant part of the process that revealed how erroneous that belief was.[2]

Public officials in ten states and the Oklahoma Territory filed antitrust suits against various affiliates of the oil combination, yet none of this litigation restored or preserved competition. Three Ohio attorneys general and various minor state officials initiated a total of thirteen suits over a period of sixteen years. Every one of these actions was either dropped or decided in favor of Standard Oil. Texas authorities succeeded in ousting the Waters-Pierce Oil Company from the state after a lengthy legal battle, but United States Senator Joseph A. Bailey promptly engineered the company's readmission. Nine years later the Texas attorney general forced Waters-Pierce and two other Standard subsidiaries into receivership, only to allow Standard interests to repurchase these firms and continue operating within the state. When Tennessee expelled oil combination marketing affiliate Standard of Kentucky, the oil barons reassigned marketing responsibilities in that state to another subsidiary—Standard of Louisiana. In Kansas, the state attorney general capped a five-year antitrust campaign against three Standard companies by negotiating an out-of-court settlement that guaranteed those firms continued domination of the local petroleum industry. The Missouri Supreme Court ousted both Waters-Pierce and Standard of Indiana in 1908, but the judges subsequently reconsidered and allowed both firms to remain in the state.

The federal antitrust effort against Standard Oil of New Jersey fared no better. The suit began as an integral component of Theodore Roosevelt's policy of maximum exploitation of public hostility against the oil combination. But Roosevelt thought so little of the antitrust process that he would have settled the whole matter out of court if Standard Oil leaders had not broken off negotiations. The slow-moving federal investigation, however, eventually produced an irrefutable mountain of evidence that guaranteed Jersey Standard's conviction. Despite the strength of his case, special prosecutor Frank Kellogg requested remedies— subsequently approved by the Taft administration and further weakened by both the federal circuit court and the United States Supreme Court—that were inadequate to achieve meaningful competition in the petroleum industry. The Standard companies thus were able to operate as a closely coordinated unit for at least fifteen years after the decree took effect.

Contemporary defenders of the corporate giants had a ready explanation for the difficulties that plagued would-be trustbusters. They believed that combination was the irresistible wave of the future, a necessary prerequisite for both increased productivity and technological advance. Antitrust prosecutors habitually failed because they naively attempted to arrest the natural course of economic development and return the nation to the age of the family farm and the independent artisan.[3]

This explanation, however, simply did not apply to the petroleum industry in the first third of the twentieth century. Fundamental economic forces in that sector of the economy favored competition, not combination. The petroleum business expanded rapidly. New oil fields appeared in widely scattered sections of the country. Flush fields periodically flooded the market with vast quantities of crude. Technological innovations like the automobile created new demands for petroleum products. These factors offered oil firms outside the Standard combination great opportunities, and many of these independent companies grew rapidly several years before the Supreme Court decision against Jersey Standard in 1911. The oil trust, in spite of its vast power and unscrupulous competitive practices, could not halt the growth of these new rivals. Hence the task of antitrust litigation as it related to the oil industry was to break down artificial barriers to competition. But the suits against the Standard combination proved so ineffectual that they failed even this modest assignment.[4]

The Standard Oil cases demonstrated the grave procedural deficiencies of antitrust. They also displayed the power of an economic titan like Standard Oil to subvert this extremely delicate legal process. In addition, the federal case against Standard Oil of New Jersey provided Chief Justice White a prominent platform from which to announce the rule of reason, and the court decree in that case ordering the restructuring of the Standard empire established the pattern of cooperative oligopoly that has characterized the petroleum industry for the last sixty years.

More significant than any of these considerations, the Standard Oil cases marked the coming of age of antitrust as a powerful and popular public relations tool. At the time of this litigation, oil was growing more important to the national economy every day. Henry D. Lloyd and Ida Tarbell had already established Standard Oil as the embodiment of malevolent monopoly. President Roosevelt heightened public hostility toward the oil combination, and in this pursuit President Taft followed in the wake of his formidable predecessor. The suits against Standard Oil, which spanned two full decades and received considerable publicity, spread awareness of the antitrust process, but the public lost sight of the lack of success of this enterprise in the limitless morass of legal complications and the passage of time. The politicians who initiated these suits were not anxious to draw attention to their failures.

The antitrust crusade against Standard Oil created a double legacy. These cases helped institutionalize the antitrust suit as the politician's recourse when corporate abuse offended too deeply the public's collective sense of propriety. The suits also created the powerful precedent that prosecutors disregard the futility that generally characterized antitrust litigation. Theodore Roosevelt deserves special acknowledgment as one of the earliest, most skillful, and certainly most influential practitioners of this emerging form of political deception. Roosevelt did not believe in the ultimate goals of antitrust, yet he increased the tempo of federal antitrust activity and began a case against the nation's most despised monopoly. His administration's pursuit of the oil trust was marked with indifference and irresolution, but his public relations campaign against Standard Oil was relentless. In antitrust policy, Roosevelt discovered that it was far more important to appear to take a firm stance against corporate abuse than to emerge victorious from an extended struggle through the courts. Other politicians have adhered to this same principle.

In September 1939 Adolf Hitler's army attacked Poland and began the European phase of World War II. Fifteen months later the Japanese raid on Pearl Harbor brought the United States into the conflict. In the intervening period the American public grew increasingly concerned over cartel agreements and patent sharing arrangements between major German corporations such as Krupp and I. G. Farben and several leading American firms, including General Electric, DuPont, and Standard Oil of New Jersey. In several instances, these contracts—which involved such vital industries as chemicals, optical instruments, magnesium, aviation fuel, and synthetic rubber—both increased German war production and impeded America's initial attempt to achieve a state of economic preparedness. By early 1941, this situation had drawn critical scrutiny from both liberal and conservative quarters in the press.[5]

Thurman Arnold, who in 1938 became head of the Antitrust Division of the Justice Department, was an advocate of vigorous antitrust enforcement. He expanded the division staff and dramatically stepped up the pace of federal antitrust litigation. In 1940 he asserted that the war in Europe made the antitrust suit a more important tool than ever before; it could stop wartime profiteering, protect independent businessmen, and stimulate greater defense production. In pursuit of these goals, Arnold launched investigations into the relationships between American corporations and German firms.[6] The link between Standard Oil of New Jersey and I. G. Farben figured prominently in these probes.

Federal investigators discovered that Jersey Standard began its secret association with I. G. Farben in the late 1920s. In March 1926, a Standard official visited the Farben laboratories in Mannheim and witnessed a new process to produce synthetic gasoline from Germany's abundant supply of coal. The Standard directors feared that this technology might eventually give Farben the power to upset the world oil market. Therefore, in November 1929, they concluded a series of agreements with the German firm to foreclose that possibility. Jersey Standard gave I. G. Farben 546,000 shares of its stock (2 percent of the total) valued at $36 million in exchange for Farben's pledge to stay out of the oil business beyond the borders of Germany. Standard Oil promised not to enter the world chemical industry, even in the United States. In addition, the companies agreed to exchange freely patents and research in areas relating to both petroleum and chemicals.[7]

The Standard-Farben connection eventually involved products of major military significance. Synthetic rubber, one of the most important, allowed the Wehrmacht to carry on mechanized war independent of the natural rubber supplies of the Far East that were geographically beyond the reach of Germany. Although Standard acquired some important technical data from Farben, the exchange of information between the two companies during the 1930s consistently worked to the advantage of the Third Reich and against the national interest of the United States.

In 1932 I. G. Farben successfully completed a pilot project to manufacture an important synthetic, Buna rubber, and two years later began large-scale manufacture of it. Standard officials regarded the process as falling within the bounds of their research-sharing agreement with Farben, and they tried repeatedly throughout the 1930s to get technical information and patent rights to this new product. But the German government prevented I. G. Farben from releasing the information. Nazi officials considered the synthetic a strategic material and refused to authorize its development outside Germany. In 1937 Frank Howard, the Standard official in charge of negotiations with Farben, informed his firm's executive committee that information on Buna "has not been forthcoming as a result of the German Government's refusal, because of military expediency, to permit I. G. to reveal such information to anyone outside Germany."[8]

That same year Jersey Standard technicians developed another synthetic, Butyl rubber. Despite Farben's consistent refusal to turn over information on Buna and despite the strategic implications of the new product, Standard sent Farben samples and complete technical data on Butyl in March 1938. At that time, Frank Howard wrote: "I am convinced that it is not only the right thing to do, but the best thing from every standpoint to pass on to them full information on the copolymer [Butyl] at this time. I do not believe we have anything to lose by this which is comparable with the possible benefit to our interests."[9] Apparently Howard thought that releasing full technical data on Butyl would help Farben get permission from Nazi officials to release similar information on Buna. And in fact, Standard pressed hard, although without success, for this information in the months immediately following the release of Butyl.[10]

In mid-1939 Standard obtained permission from Farben under the terms of the patent-sharing agreement to contract with American rubber companies to begin commercial development of the Standard-created synthetic Butyl. But Frank Howard was so anxious not to strain relations with Farben that he recommended postponing discussions with the rubber manufacturers for several months. In spite of the opposition of other ranking Standard officials, Howard's opinion prevailed, and Jersey Standard did not release Butyl to the American rubber industry until September 1940, several months after Hitler's army had conquered France.[11]

The outbreak of war in Europe made obsolete Farben's policy of restricting development of Buna by withholding patent rights. Countries at war with Germany were no longer obligated to respect Farben's legal claims. Therefore, on September 25, 1939, Jersey Standard and Farben signed a new agreement at the Hague designed to meet wartime conditions. Farben finally released to Standard Oil the patent rights for Buna in the United States, the British and French empires, and Iraq, but the Germans still refused to provide the pertinent technical information so the patent release was merely an authorization to experiment in this field. Nevertheless, the new agreement required that both parties periodically exchange reports on revenue from Buna and share profits "in such a manner as may seem most fair and advantageous." The Standard licensing policy for Buna in the United States featured unusually high royalties and required the licensee to surrender any improvements to the Buna process to Farben and Standard Oil.[12]

In the closing months of 1941, while the Justice Department methodically concluded its year-long investigation of the Standard-Farben agreements, diplomatic relations between the United States and Japan rapidly deteriorated. Federal officials now realized that a rubber shortage with serious military implications was probable. A Japanese attack could easily cut off American natural rubber supplies from the Far East, and the synthetic rubber industry in the United States was seriously underdeveloped for reasons that were all too apparent to the Antitrust Division. In a last-minute attempt to deal with this problem, the government filed suit against Jersey Standard in November 1941, charging that agreements with I.G. Farben violated the Sherman Antitrust Act. The suit sought to dissolve the Standard-Farben patents pertaining to synthetic rubber and several other important technical fields.[13]

The suit was quickly settled. Jersey Standard officials pleaded nolo contendere, paid a token fine of $50,000, and signed a consent decree on March 25, 1942, that released patents for both Buna and Butyl to all rubber manufacturers for use during the wartime emergency without royalty payments. After that time the courts were to fix reasonable royalties.[14] Despite their plea in the case, Standard officials released a statement to the press professing total innocence: "The company realizes that to obtain vindication by trying the issue in the courts would involve months of time and energy of its officers and many of its employees. Its war work is more important than court vindication."[15]

The weakness of the court-imposed remedies may have rivaled Standard's recently discovered patriotism as an inducement to sign the decree. The fine of $50,000 was trivial to a company of Jersey Standard's size. In fact, Thurman Arnold later admitted that the size of the firm was "somewhat out of proportion" with fines in comparable cases.[16] Even without the decree the Farben-Standard patents undoubtedly would have been made available during the war. Furthermore, Farben still refused to release full technical information on Buna. Finally, Arnold himself noted that he "was reluctant to sign the decree" because it lacked a provision allowing court supervision of future relations between Jersey Standard and Farben.[17] In view of the Hague agreement of 1939, Arnold considered future judicial oversight essential.

Arnold, a strong antitrust advocate, went along with this settlement because of the war: "Of course, I would not have given my consent to this decree if I had not felt that, considering all the pressures which are upon us today, the decree would be a substantial contribution to the war program."[18] The full cooperation of Standard Oil of New Jersey was essential to a successful military effort. Antitrust litigation repeatedly had proven to be a time-consuming and generally ineffectual process. Both Jersey Standard and the synthetic rubber patents were too important to the war economy to be tied up in court during a national emergency.

Although Arnold could not pursue the case against Jersey Standard in court, he was able to maximize the public relations potential of the suit. Senator Harry S. Truman's Committee on National Defense gave Arnold the oportunity to underscore the role of the Antitrust Division in dealing with the synthetic rubber shortage. The committee, which had been established in 1941 to investigate the nation's defense capabilities in the event of war, began to hold hearings on several vital war industries shortly after Pearl Harbor. It opened public hearings on the rubber situation on March 5, 1942. The early testimony revealed that natural rubber supplies were dangerously low and that synthetics were not available to compensate for the shortage.[19]

On March 26, the day after the antitrust suit was settled by the consent decree, Thurman Arnold appeared before the Truman committee. The assistant attorney general reviewed Jersey Standard's long association with Farben and then asserted that the "cartel arrangements with Germany . . . are the principal cause of the shortage of synthetic rubber."[20] Arnold specifically stated that Standard's motives were economic. But Senator Truman, after hearing Arnold's testimony, was less inclined to give Standard the benefit of the doubt. When asked if the Hague agreement of 1939 was treasonable, he replied: "Why yes, what else is it?"[21]

The hearings provided front-page news all over the country, and the press reported Arnold's sensational charges in great detail. Both *Time* and *Newsweek* gave the proceedings full coverage, with *Time* observing: "Seldom has a U.S. business firm taken such a smearing as the Standard Oil Co. of New Jersey got last week." Some liberal accounts, following the lead of Senator Truman, portrayed Standard Oil's behavior as treasonable. The *New Republic* ran an article, "Standard Oil: Axis Ally," that ominously proclaimed, "Sooner or later businessmen who ally themselves with fascism become fascists." *PM*, a New York newspaper, ran a series of open letters of John D. Rockefeller, Jr., urging him and other stockholders to change the policies of Standard Oil. The paper charged that under the current board of directors, Jersey Standard had become an "ally of Hitler, an economic enemy agent" operating within the United States.[22]

But Arnold overstated his case when he placed the complete responsibility for the synthetic rubber shortage on Standard Oil. The Senate committee heard other testimony demonstrating that the federal government had declined on several occasions to accelerate research and development in synthetic rubber in the months before Pearl Harbor.[23] Arnold also greatly exaggerated the significance of the Antitrust Division's work in this matter. The Standard-Farben relationship went unchallenged for well over a decade, the antitrust investigation consumed more than a year of critical time, and the decree accomplished little of substance. In fact, some commentators theorized that the highly publicized attack on Standard Oil was meant to divert attention from governmental blundering.[24] The rubber shortage certainly placed the Roosevelt administration in an embarrassing position, and Arnold's testimony created the impression of strong federal action to deal with this serious national security problem. Arnold's behavior before the committee demonstrated that he, like earlier trustbusters, had learned that in the field of antitrust massive publicity and the appearance of activity were far more important than the actual results of a particular suit.

Trust busting went out of style during World War II, but by the late 1940s the oil companies had again drawn public attention to themselves. In 1948 a Senate investigation disclosed that the Arabian-American Oil Company—a joint operation in Saudi Arabia composed of Standard of California, Texaco, Jersey Standard, and Standard of New York—had sold oil to France and Uruguay for $1.00 or less per barrel, while it sold oil to the United States Navy for $1.23. Shortly thereafter the European Cooperation Administration, which was supervising the rebuilding of Europe under the Marshall Plan, discovered that the oil companies were charging it substantially more than other customers. In the spring of 1949 a sudden increase in the price of gasoline by Jersey Standard led to widespread public criticism and an investigation by the Senate Committee on Banking and Currency. Amid these various complaints, Justice Department investigators again began rummaging through oil company files.[25]

The Federal Trade Commission, not the Justice Department, scored the most sensational blow against big oil in the early postwar period. In 1952 the commission completed a study of the international operations of the major American oil companies. The report, *International Petroleum Cartel,* charged that the world oil business was controlled by only seven firms, three of them former members of Rockefeller's old oil trust: Standard Oil of New Jersey (now known as Exxon), Standard Oil of New York (Mobil), and Standard Oil of California (Chevron). Texaco, Gulf Oil, British-based Shell Oil and British Petroleum were the other members. Beginning in the late 1920s, these firms had evolved a set of agreements to restrict competition. These agreements included mutual control of all major oil-producing fields outside the United States, control of all foreign refining operations, limiting excess production, control of petroleum technology, and division of world oil markets. Although Jersey Standard, which acted as the leader of the American companies, claimed that it had withdrawn from these agreements in 1938, the practices remained the custom of the trade.[26]

The Truman administration released the contents of this report in a manner designed to maximize the political impact of the relevations. The Federal Trade Commission disclosed in March 1952 that it had just completed this highly damaging study. It at first classified the report top secret, insisting that the information would support Soviet propaganda in the Middle East if it were made public. But shortly after the Democratic presidential nominating convention, Senator John Sparkman, chairman of the Small Business Committee and the party's vice-presidential nominee, requested that the government release the report. Truman then promptly declassified it, omitting about forty pages for security reasons. Senator Sparkman made the most of this opportunity. In late August his Small Business Committee put its name on the study (though it had done none of the work), and Sparkman wrote a fiery introduction emphasizing the nation's efforts "to hold in check the power of giant organizations." The Justice Department then underscored the Democratic administration's hostility to big oil by taking legal action to recover the $67 million that the oil companies had overcharged the European Cooperation Administration.[27]

With public passions enflamed over oil company abuses, the Truman administration judged the time right for an antitrust suit. The president requested that the Justice Department assemble a federal grand jury to look into possible criminal indictments of the oil companies. It convened in Washington on September 2, 1952, barely a month before the presidential election.[28] The sensational disclosures about the worldwide oil cartel and the blatant political motives behind the antitrust suit vied with each other for the attentions of the press. *Time* labeled the whole affair a "peep show" and likened the Truman administration to a "deft stripteaser" who had been "peeling off just enough gossamer to give the customers some tantalizing glimpse of the 'secret oil cartel.'"[29] Even *Fortune* magazine had to admit that the multinational oil companies were an obvious target for the Antitrust Division: "Even if government trust busters did not have the stimulus of a presidential election, they could hardly be expected to overlook the foreign operations of American oil companies. When in recent times have they been presented with material that lent itself to such easy and dramatic conclusions?"[30]

Just as the Justice Department was gearing up for another assault on the oil companies, events halfway around the world intervened. By the early 1950s Iran had become a major petroleum-producing region; one company, British Petroleum (then known as the Anglo-Persian Oil Company), controlled the country's entire oil industry. Because they played such a vital role in the economy of the region, the leaders of British Petroleum viewed themselves as agents of social and economic progress, but they were particularly slow in sensing the changing mood of the third world. The Iranians, on the other hand, wanted genuine independence and increasingly resented the dominant position of the oil company. The fact that one foreign corporation controlled Iran's most important industry made British Petroleum a convenient scapegoat for all the country's problems. In 1951 a new leader, Muhammed Mossadeq, rose to prominence by calling for nationalization of Iran's oil industry. The Iranian prime minister was murdered after he insisted that the country could not legally repudiate its agreement with British Petroleum. After six weeks of strikes and riots, Mossadeq became prime minister, and the Iranian legislature promptly nationalized the oil fields.[31]

The Iranian action was a direct challenge to Great Britain because that government owned half of British Petroleum, but after an initial period of confusion, British officials decided against direct intervention. Instead, British Petroleum enlisted the support of the major international oil companies in a worldwide boycott of Iranian crude, a successful strategy because of an oil glut. British Petroleum increased its production in other countries, particularly Kuwait, to make up the loss. The other major companies, which also had a substantial interest in demonstrating to Iran and the other producing nations that nationalization would not work, followed suit. With this policy in place, big oil and the British government waited for the loss of oil revenue to bring down the Mossadeq regime.[32]

By late 1952 Dean Acheson, the American secretary of state, decided that something must be done to break the long stalemate over Iranian oil. Iran bordered on the Soviet Union, and he worried that it might be pushed into the Communist orbit if the impasse continued indefinitely. Acheson concluded that an American initiative was in order, but he needed the support of the American oil companies, which controlled the facilities to bring Iranian oil back into the world market. Before he could gain the cooperation of the companies, the antitrust crusade against the oil industry would have to be disposed of.[33]

The inevitable clash between Secretary of State Acheson and the Antitrust Division is outlined in reports submitted to the National Security Council during the final days of the Truman administration. The State Department, backed by Defense and Interior, urged that the government adopt a policy of supporting the international operations of big oil. Their report emphasized that petroleum had become the lifeblood of the economy of the industrial West and that working with the companies was the only way to guarantee an abundant and uninterrupted supply of oil. Criminal charges under the Sherman Act in the United States would expose these companies to legal harassment in countries around the world. Even worse, it would strengthen the hand of advocates of nationalization in Latin America and the Middle East. Finally, because of the global ideological struggle with the Soviet Union, the American government could not "afford to leave unchallenged the assertion that these companies are engaged in a criminal conspiracy for the purpose of predatory exploitation." Thus it was imperative that the government not file criminal charges and seek remedies through a less drastic civil suit, which might be settled out of court through a consent decree.[34]

The Justice Department responded that the multinational oil companies had damaged national security. The monopoly power of big oil must be broken because petroleum was so critical to national defense and the world economy. Agreements made by the multinationals "are in effect private treaties negotiated by private companies to whom the profit incentive is paramount. The national security should rest instead upon decisions made by the Government with primary concern for the national interest." The Justice Department also noted: "We cannot promote free enterprise and productivity abroad unless we are seen to conscientiously enforce our laws designed to preserve them for our own economy." Finally, it stressed the importance of proceeding with a criminal suit. A civil suit would take about six years to complete and at best would end in a decree to restructure the industry that had little chance of success. A criminal suit was a far better route to meaningful change.[35]

Six days after these reports were submitted, Truman directed his attorney general to file a civil suit against the oil companies. The election over, the Truman administration decided that the foreign policy complications were too great to justify pursuing this matter. In truth, the antitrust record of accomplishment was and is so meager that it is hard to fault Truman for backing down. Nevertheless, the new Eisenhower administration decided to press on with the civil case that the Justice Department already had said could not be won. The new Republican president was just as interested as Truman in paying lip-service to the time-worn clichés of competition and free enterprise. On April 21, 1953, the Justice Department filed suit against Standard Oil of New Jersey, Standard Oil of New York, Standard Oil of California, Texaco, and Gulf, charging them with operating for twenty years within a worldwide system designed to control production, refining, research, and transportation of oil. The Justice Department then plunged into an investigation that occupied it for the duration of Eisenhower's two terms in the White House and beyond.[36]

With the Justice Department out of the way, Eisenhower's secretary of state, John Foster Dulles, moved decisively. In August 1953 the Central Intelligence Agency, headed by the secretary of state's brother, Allan Dulles, provided the funds, equipment, and tactical support to overthrow the uncooperative Mossadeq. General Zahedi then took over the government, and the shah, who had earlier fled to Europe, returned to Iran.[37]

Secretary Dulles then dispatched Herbert Hoover, Jr., to Teheran to negotiate a settlement of the oil dispute with the Iranian government. The new international consortium erected to replace the old British Petroleum monopoly granted British Petroleum 40 percent of the crude oil from Iran. The five American oil companies named in the antitrust suit shared another 40 percent, while Shell received 14 percent, and CFP, a French company, took the remaining 6 percent. Profits were to be divided equally between Iran and the oil companies. In order to prevent worldwide overproduction, the companies devised a complex system to restrict crude output in Iran to levels required by the least demanding of the partners. This agreement was kept secret from both the shah and the American public for twenty years.[38]

A consortium that carved up the Iranian petroleum industry and limited production made a farce out of the antitrust suit against the oil cartel, nevertheless the Justice Department continued to assemble evidence in the case for over eight years. The cases against Jersey Standard, Texaco, and Gulf were finally settled through consent decrees negotiated between 1960 and 1963. In 1968 charges were dropped against Standard of California and Standard of New York. The decrees contained the customary prohibitions against price fixing, market allocations, and the like, but the Justice Department built into the judgments loopholes large enough to permit the continued operation of the Iranian consortium and similar joint international operations. As a Justice Department official put it: "The decrees provide expressly that the companies are not prohibited—that is, prohibited by the decrees—from participating in joint production operations, joint refining operations, joint pipeline operations or joint storage in foreign nations." After ten years of investigation, these terms hardly amounted to a great victory for the government.[39]

The antitrust suit of 1953 against the five American participants in the international oil cartel followed a pattern that has been repeated many times since the passage of the Sherman Act in 1890. Economic abuses and sensational disclosures led to public outrage against big oil. An antitrust suit followed at just the right time to create the maximum public impression of strong government action. The government then promptly abandoned energetic pursuit of the case but allowed the issue to drag on through the courts until the public had grown apathetic. The case was finally settled in a fashion that left the anticompetitive practices of the oil companies untouched and did nothing but relieve the government of the necessity of taking real action.

In September 1969 Colonel Muammar Qadaffi and a group of young army officers overthrew the corrupt regime of King Idris and seized control of oil-rich Libya. This North African country was a key source of petroleum for Western Europe, and Qadaffi intended to use oil to increase his nation's wealth dramatically and to wage ideological war against Israel. In mid-1970 Qadaffi demanded a fifty cent per barrel raise in the price of Libyan oil. After some resistance from the major oil companies, he got most of what he wanted. Qadaffi's success had an electrifying effect on other oil-producing countries. Iraq, Algeria, Kuwait, and Iran all quickly raised the tax or the profits of oil companies. In December 1970 the members of the Organization of Petroleum Exporting Countries met in Caracus, Venezuela, intent on extracting better terms from the oil companies. They passed resolutions calling for higher taxes on oil companies and high prices for their crude. They also resolved to work together toward these goals at a new round of negotiations with the companies in Teheran, Iran, early in 1971.[40]

The leaders of the major oil companies frantically attempted to form a common front against the increasing militant demands of OPEC. In January 1971 representatives of Exxon, Mobil, Chevron, Texaco, Gulf, Shell, British Petroleum, and sixteen other oil companies met in the New York offices of John McCloy, a lawyer who had represented many of the major oil companies in antitrust and other matters for well over a decade. During a three-day marathon session the oil men hammered out two agreements. First, they insisted on dealing with OPEC as a single unit: "We cannot further negotiate the development of claims by member countries of OPEC on any other basis than one which reaches a settlement simultaneously with all producing governments." Second, they devised the Libyan producers' agreement. They promised not to negotiate with the Libyan government without the knowledge and approval of the other companies. In addition, the companies agreed that if the Libyan government ordered one of the companies to cut back its oil production, all of the companies would share the cutback in predetermined proportions.[41]

These agreements completely contradicted the principles of the Sherman Act. Nevertheless, John McCloy was able to work around the antitrust problem by obtaining a business review letter from the Antitrust Division. The Justice Department cannot provide legal advice to private companies concerning the antitrust implications of a proposed joint business venture, and it lacks the authority to exempt a particular anticompetitive practice from the provisions of the Sherman Act, but it can indicate its attitude toward a particular joint venture through a business review letter. Although the document is not legally binding, it does provide informal governmental approval of a questionable business arrangement. In this particular instance the Justice Department concluded that national security required that the oil companies band together to form a "counter-vailing force" against OPEC. On February 14, 1971, an agreement was signed at Teheran that granted OPEC a thirty-cent per barrel price increase. A few weeks later Libya agreed to a seventy-six-cent increase at Tripoli. McCloy considered these relatively modest in-creases a solid justification for joint action.[42]

After the original business review letter in 1971, the oil com-panies continued to work in unison and continued to receive sup-plementary clearances from the Justice Department. In October 1973 OPEC announced its intention to reopen the entire price ques-tion at a Vienna conference. The oil companies asked that their an-titrust clearance be extended once again to cover this new round of negotiations, but because of the wide range of issues to be covered at the Vienna talks, the Justice Department decided to reexamine the whole issue of cooperation between the oil companies. After in-terviewing leading officials of fourteen separate companies and ex-amining company records, it concluded that the joint action had helped stabilize oil prices and ensured adequate supplies. The Justice Department therefore decided to extend the business review clearance.[43]

The Vienna negotiations were a disaster. The meeting began under the intense pressure of the Arab-Israeli war. When the oil companies balked at a posted price of $5 per barrel, the Arabs angrily walked out. They reconvened in December in Teheran and unilaterally set the price at $11.65 a barrel, a quadrupling in four months.[44]

In all, the Justice Department granted the oil companies six separate antitrust clearances between 1971 to 1973. The policy was a catastrophic failure. The Justice Department rationalized their subversion of the antitrust law by asserting the need to establish a countervailing force to OPEC. Yet it can be reasonably argued that the oil companies were too confident and that their intransigence at key moments contributed to the price explosion. It is hardly sur-prising that both the Justice and State departments went to great lengths to keep the business review letters secret. When the Senate's Subcommittee on Antitrust requested information relating to an-titrust and the oil companies in early 1974, the Justice Department refused to turn over the information, citing national security.[45]

When public anger focused on big oil as price hikes followed shortages in the winter of 1973-1974, government officials resorted to the time-honored tradition of investigation in an attempt to placate public opinion. Senator Henry Jackson of Washington, chairman of the Permanent Subcommittee on Investigations and a leading presidential aspirant, was both well placed and well motivated to lead the attack. He summoned the chiefs of the major oil companies in Washington and grilled them in front of television cameras concerning their enormous profits. The Jackson hearings ended without producing much new information about the industry. But another Senate subcommittee carried out the more tedious work of uncovering the details concerning the international oil companies. In 1974 the Subcommittee on Multinational Corporations, chaired by Senator Frank Church of Idaho, amassed the most important body of evidence on the oil companies since the Federal Trade Commission report of 1952. The committee subpoenaed and published the secret business review letters of 1971-1973, as well as a host of hitherto secret documents.[46]

Antitrust remained a tempting last resort for embarrassed officials when the public became aware of the government's failure to control big business. At the multinational hearings Thomas Kauper, the head of the Antitrust Division, spoke confidently of the serviceability of the Sherman Act even as he revealed how the Justice Department had secretly approved big oil's noncompetitive negotiations with OPEC. Kauper informed the committee that an "energy unit" had been recently established within the Antitrust Division of the Justice Department. This section currently provided employment for eight bureaucrats, six attorneys, and two economists and shortly would be expanded. Before the committee, Kauper underscored the Justice Department's role as "advocates for the free market system" and declared, "We believe that the United States Government should take a more active role to insure protection for the interests of consumers as well as [oil] producing companies."[47] Senator Church, even after the disclosure of the business review letters, also made the customary references to the aging Sherman Act: "I cannot think of an industry that needs a more thorough or penetrating review when it comes to the antitrust laws of this country than the oil industry, and I think that this should have top priority within the Department."[48] The report issued by the committee called on the Antitrust Division to disrupt the comfortable cooperative system of the multinational oil companies.

The politicians once again appeared to be falling into the familiar rut. An outburst of public indignation against big oil had been followed by a series of investigations and the inevitable calls for an antitrust suit. But this time the Justice Department did not rush into action. The customary procedural problems with antitrust remained, and the global operations of the major oil companies both limited the Antitrust Division's jurisdiction and introduced complex national security considerations into any antitrust action. The difficulties involved in antitrust had grown so enormous that they appeared to outweigh the public-relations benefits of filing suit. Perhaps the politicians have finally moved beyond the tradition, pioneered by Theodore Roosevelt in the Standard Oil case of 1911, of filing and forgetting antitrust cases against big oil. In fact, the abandonment of antitrust as a tool against Exxon, Mobil, Chevron, and others would be a good first step in developing some realistic scheme for coping with the power of the descendants of John D. Rockefeller's oil monopoly. Antitrust has never worked, and it has been used repeatedly to obscure the inability of the federal government to protect the public from the excesses of big business.

Notes

INTRODUCTION

1. On the emergence of antitrust as a political issue, see William Letwein, *Law and Economic Policy in America* (New York: Random House, 1965), pp.54–70, and Hans B. Thorelli, *The Federal Antitrust Policy* (Baltimore: Johns Hopkins Press, 1955), pp. 137–43. For a quantititative survey of public opinion on the trusts during the 1880s see Louis Galambos, *The Public Image of Big Business in America, 1880–1940* (Baltimore: Johns Hopkins Press, 1975), pp. 47–48. Thomas H. McKee, *The National Conventions and Platforms of All Political Parties* (Baltimore: Friedwald Company, 1906), pp. 241–47, reprints all the pertinent party platforms.

2. Ralph Hidy and Muriel Hidy, *Pioneering in Big Business* (New York: Harper and Brothers, 1955), pp. 40–49; Allan Nevins, *Study in Power: John D. Rockefeller* (New York: Charles Scribner's Sons, 1953), 1:382–402.

3. "The Fortune 500," *Fortune* 95 (May 1977): 366-367.

4. Letwin, *Law and Economic Policy,* pp. 69–70; Thorelli, *Federal Antitrust Policy,* pp. 76-85; Henry R. Seager and Charles B. Gulick, *Trust and Corporation Problems* (New York: Harper and Brothers, 1929), pp. 49-60; Eliot Jones, *The Trust Problem in the United States* (New York: Macmillan Company, 1926), p. 21. Gabriel Kolko, *The Triumph of Conservatism* (Chicago: Quadrangle Books, 1967), chaps. 1, 2, discusses the difficulty of achieving a monopoly in the chaotic economy of the United States.

5. *Central Ohio Salt Company* v. *Guthrie,* 35 Ohio 666, 667 (1880). Some examples of the successful use of the common law to limit monopolies are: *Mallory* v. *Hanaur Oil-Works,* 86 Tenn. 598 (1888); *State* v. *Nebraska Distilling Company,* 29 Neb. 700 (1890); *People* v. *North River Sugar Refining Company,* 121 N.Y. 582 (1890).

6. Seager and Gulick, *Trust and Corporation Problems,* pp. 342-47; also see Federal Trade Commission, *Trust Laws and Unfair Competition* (Washington: Government Printing Office, 1916), for a comprehensive survey of the early state antitrust laws.

7. Seager and Gulick, *Trust and Corporation Problems,* p. 348.

8. Thorelli, *Federal Antitrust Policy,* pp. 164-232, is the fullest account of the legislative origins of the Sherman Act.

9. For example, see Senator Sherman's statements in *Congressional Record* 21, pt. 3, 51st Cong., 1st sess., p. 2460. Regarding public opinion and the passage of the Sherman Act, Louis Galambos concluded: "If the legislators calculated that a vague measure would relieve some tensions accumulating around the trust issue and mollify that part of public opinion which was aroused, my study indicates that they were excellent judges of their constituents' frame of mind." *Public Image of Big Business,* p. 78.

10. Act of July 2, 1890, 26 U.S. Stat. 209, 51st Cong., 1st sess.

11. *Congressional Record* 21, pt. 4, 51st Cong., 1st sess., p. 3146.

12. Thorelli, *Federal Antitrust Policy,* pp. 201, 210.

13. 26 U.S. Stat. 209.

14. Walton Hamilton, *The Pattern of Competition* (New York: Columbia University Press, 1940), chap. 3, is the best discussion of procedural problems in antitrust litigation.

15. Quoted in Matthew Josephson, *The Politicos* (New York: Harcourt Brace, 1938), p. 460. Josephson charged that the law was intentionally vague. This view also appears in Charles A. Beard and Mary R. Beard, *The Rise of American Civilization,* rev. ed.(New York: Macmillan Company, 1930), 2:327, and Samuel E. Morison and Henry Steele Commager, *The Growth of the American Republic,* 4th ed. (New York: Oxford University Press, 1951), 2:144.

16. Letwin, *Law and Economic Policy,* pp. 95-99, and Thorelli, *Federal Antitrust Policy,* pp. 225-32.

17. Richard Hofstader, "What Happened to the Antitrust Movement?" in his *The Paranoid Style in American Politics* (New York: Vintage Books, 1967), pp. 188-237, is a good introduction to the evolution of modern antitrust.

CHAPTER 1

1. Ida Tarbell, *History of the Standard Oil Company* (New York: McClure, Phillips and Company, 1904), 1: chap. 2; Allan Nevins, *Study in*

Power: John D. Rockefeller (New York: Charles Scribner's Sons, 1953), 1: chaps. 2–5.

2. Tarbell, *History of the Standard Oil Company,* 1: chaps. 2-3; Nevins, *Study in Power,* 1: chaps. 6-9, 13; Ralph Andreano, "The Emergence of New Competition in the American Petroleum Industry Before 1911" (Ph.D. diss., Northwestern University, 1960), p. 6.

3. Nevins, *Study in Power,* 1: 382-87.

4. Ralph Hidy and Muriel Hidy, *Pioneering in Big Business* (New York: Harper and Brothers, 1955), p. 44.

5. Nevins, *Study in Power,* 1: 397.

6. Hidy and Hidy, *Pioneering in Big Business,* p. 48.

7. This treatise was William W. Cook's *Trusts: The Recent Combinations in Trade* (New York: L.K. Strouse and Company, 1888), appendix B, pp. 78–89.

8. Tarbell, *History of the Standard Oil Company,* 2: 141-43.

9. Hoyt Landon Warner, *Progressivism in Ohio* (Columbus: Ohio State University Press, 1964); Philip D. Jordan, *Ohio Comes of Age* (Columbus: Ohio State Archaeological and Historical Society, 1943); and E. H. Roseboom and E.P. Weisenburger, *A History of Ohio* (New York: Prentice-Hall, 1934), all discuss business domination of Ohio politics in the late nineteenth century. An impassioned contemporary account of corruption in Ohio is Allen O. Meyers, *Bosses and Boodle in Ohio Politics* (Cincinnati: Lyceum Publishing Company, 1895).

10. *Ohio* v. *Standard Oil Company of Ohio,* 49 Ohio 137, 140, 155, (1892).

11. Allan Nevins, *John D. Rockefeller: The Heroic Age of American Enterprise* (New York: Charles Scribner's Sons, 1940), 2: 139.

12. Warner, *Progressivism in Ohio,* p. 6.

13. David K. Watson, *History of American Coinage* (New York: G.P. Putnam's Sons, 1899), p. 239. Bibliographical material on Watson appears in George I. Reed, ed., *Bench and Bar of Ohio* (Chicago: Century Publishing Company, 1897), 2: 257-58. Also see Joseph B. Foraker, *Notes of a Busy Life* (Cincinnati: Stewart and Kidd Company, 1916), 1: 79, 111.

14. David K. Watson, *The Constitution of the United States* (Chicago: Callaghan and Company, 1910), 2: 577.

15. Quoted in Tarbell, *History of the Standard Oil Company,* 2: 143.

16. The most important decisions were: *Mallory* v. *Hanaur Oil-Works,* 88 Tenn. 598 (1888); *State* v. *Nebraska Distilling Company,* 29 Neb. 700 (1890); and *People* v. *North Sugar Refining Company,* 121 N.Y. 582 (1890).

17. Attorney General of Ohio, *Annual Report of the Attorney General of Ohio* (Columbus, 1899). Monnett failed to prove these charges. At the time he made them, he was involved in a bitter dispute with the Standard organization.

18. Quoted in Herbert Croly, *Life of Mark Hanna* (New York: Macmillan Company, 1923), p. 268.

19. Quoted in ibid., p. 269; Tarbell, *History of Standard Oil Company,* 2: 147.

20. On Choate, see Theron G. Strong, *Joseph H. Choate* (New York: Dodd, Mead and Company, 1917); Edward S. Martin, *Life of Joseph Hodges Choate* (New York: Charles Scribner's Sons, 1920); and *Dictionary of American Biography* (New York: Charles Scribner's Sons, 1928-36), 4: 83-86. For Dodd, *DAB,* 5: 341-42; and for Kline, Reed, ed., *Bench and Bar of Ohio,* 2: 305-06.

21. *Ohio* v. *Standard Oil,* at 159.

22. Ibid., at 167-68.

23. Ibid., at 184.

24. Ibid., at 165.

25. Ibid., at 185.

26. Ibid., at 186-87.

27. Section 6789 Rev. Stat. Ohio (1890).

28. *Ohio* v. *Standard Oil,* at 167. Watson cited *State* v. *Railway Company,* 40 Ohio 504 (1884).

29. *Ohio* v. *Standard Oil,* at 174.

30. Ibid., at 188.

31. Ida Tarbell, *The Nationalizing of Business* (New York: Macmillan Company, 1936), p. 208. For similar statements, see Gilbert H. Montague, *The Rise and Progress of Standard Oil* (New York: Harper and Brothers, 1904), p. 115; Hidy and Hidy, *Pioneering in Big Business,* p. 219; Nevins, *Heroic Age,* 2: 149.

32. *People* v. *North River Sugar Refining Company,* 121 N.Y. 582, 619 (1890).

33. *Bank of United States* v. *Deveaux,* 5 Cranch 61 (1809); *Booth* v. *Bunce,* 33 N.Y. 139 (1865); *Hibernia Insurance Company* v. *St. Louis and New Orleans Transit Company,* 13 Fed. 516 (1882); *Montgomery Web Company* v. *Dienelt,* 133 Penn. 585 (1890); *Interstate Telegraph Company* v. *Baltimore and Ohio Telegraph Company,* 51 Fed. 49 (1892); *Brundred* v. *Rice,* 49 Ohio 640 (1892).

34. *Ohio* v. *Standard Oil,* at 186.

35. *Salt Company* v. *Guthrie,* 35 Ohio 666 (1880); *Emery* v. *Candle*

Company, 47 Ohio 320 (1890); *Richardson* v. *Buhl,* 77 Mich. 632 (1889); *India Bagging Association* v. *B. Kock and Company,* 14 La. 168 (1859); *Morris Run Coal Company* v. *Barclay Coal Company,* 68 Penn. 173 (1871); *Raymond* v. *Leavitt,* 46 Mich. 447 (1881); *De Witt Wire-Cloth Company* v. *New Jersey Wire-Cloth Company,* 14 N.Y. Supp. 277 (1891); and *Chapin* v. *Brown Brothers,* 83 Iowa 156 (1891).

36. S.C.T. Dodd, "The Present Legal Status of Trusts," *Harvard Law Review,* 7 (October 1893): 165. Also see William F. Dana, "Monopoly under the National Anti-Trust Act," *Harvard Law Review* 7 (January 1894): 338-55; Frank J. Goodnow, "Trade Combinations at Common Law," *Political Science Quarterly* 12 (June 1897): 212-45.

37. *Ohio* v. *Standard Oil,* at 140. Italics added.

38. *State* v. *Railway Company,* 40 Ohio 504 (1884).

39. *State* v. *Nebraska Distilling Company,* 29 Neb. 700 (1890), and *People* v. *North Sugar Refining.*

40. Quoted in Tarbell, *History of Standard Oil Company,* 2: 151.

41. William T. Spear, Joseph Bradbury, Franklin Dickman, Thaddeus Minshall, and Marshall J. Williams sat on the supreme court of Ohio in 1892. Brief accounts of the lives of all members of the court are in Carrington T. Marshall, *A History of the Courts and Lawyers of Ohio* (New York: The American Historical Society, 1934), 1: 256-59, and Edgar B. Kinkead, "A Sketch of the Supreme Court of Ohio," *The Green Bag,* 7 (March 1895): 290-94, *Dictionary of American Biography* contains articles on Williams, 20: 283, and Spear, 27: 439-40. The following brief biographical notices appear in the *National Cyclopaedia of American Biography* (New York: James T. White and Company, 1898–1951): Spear, 12:116-17; Bradbury, 5: 560; Dickman, 7: 517-18; Minshall, 12: 273; Williams, 12: 119. An article on Dickman appears in Reed, ed., *Bench and Bar of Ohio,* 2: 146-50; one on Spear is in *Case and Comment* 17 (May 1911): 640.

42. Conservative decisions were: inheritance tax in *Ohio* v. *Ferris,* 53 Ohio 314 (1895); lien law in *Palmer* v. *Tingle,* 55 Ohio 423 (1896); voting reform in *Karlington* v. *Board of Control,* 60 Ohio 489 (1898); eight-hour law in *Clement Construction Company* v. *Cleveland,* 67 Ohio 197 (1902). The AFL survey was Jackson H. Ralston, *Study and Report for the American Federation of Labor upon Judicial Control over Legislatures as to Constitutional Questions,* 2d ed. (Washington: Law Reporter Printing Company, 1923), p. 91. The survey of decisions is in Ohio Constitutional Convention, *Proceedings and Debates, 1912* (Columbus: F.J. Heer Publishing Company, 1922), 2: 1092-1101.

43. *Standard Oil Company of New Jersey* v. *United States,* 221 U.S. 1 (1911), *Brief for the United States,* no. 725, 1: 56-57.

44. Ibid., pp. 57-68.

45. *Brief for the United States,* no. 725, 1: 70-72; Nevins, *Heroic Age,* 2: 344.

46. United States Industrial Commission, *Report of the Industrial Commission on Trusts and Industrial Combinations* (Washington: Government Printing Office, 1900), 1: pt. 2, p. 574.

47. *Standard Oil Trust Cases, Record* (Columbus, Ohio, 1899), George Rice testimony, pp. 323-94.

48. For statements by Monnett on the subject of trusts, see New York *World,* March 6, 1899; Frank S. Monnett, "Bryan and the Trusts: An Anti-Trust View," *Review of Reviews,* 22 (1900) 439-43; Monnett, "Transportation Franchises Always the Property of Sovereignty," *Arena* 26 (August 1901): 113-27. Monnett may have had particular animosity toward the Standard Oil combination because his father was president of the Bucyrus Gas Company, a competitor with the Standard affiliate, Northwestern Ohio Natural Gas Company. Carrington T. Marshall, *A History of Courts and Lawyers of Ohio* (New York: American Historical Society, 1934), 4: 89.

49. For statistics on increased consolidation of industry from 1897 to 1903, see Thorelli, *Federal Antitrust Policy,* p. 275. Regarding increased public interest in the trust problem, Thorelli notes (on pp. 239–40) that the Library of Congress bibliography of articles dealing with the trust problem lists 25 items for the 1891–1896 period, 20 for 1897, 17 for 1898, 76 for 1899, and 105 for 1900. Listings under *Trusts* in the *New York Tribune* Index contain 2 references for 1891, 11 for 1895, and 123 for 1899.

50. Monnett indicated his political ambition by actively seeking the Republican gubernatorial nomination in 1899. Everett Walters, *Joseph Benson Foraker* (Columbus: Ohio History Press, 1948), p. 175.

51. *Standard Oil Trust Cases, Record,* George Rice testimony, pp. 388-89.

52. Ohio Senate, *Trust Investigation* (Columbus, 1898). Rice's testimony at pp. 8–22. Also see Andreano, "Emergence of New Competition," pp. 20, 38, 41-42.

53. Ohio Stat. 1898, p. 143.

54. *Standard Oil Trust Cases, Record,* "Answers to Interrogatories," pp. 454-92.

55. New York *World,* October 12, 1899; *Standard Oil Trust Cases, Record,* John D. Rockefeller testimony, pp. 435-37; *New York Tribune,* October 13, 1899, gives a list of these certificate holders.

56. *Standard Oil Trust Cases, Record,* pp. 2-99.

57. *Weekly Law Bulletin,* 41 (April 17, 1899): 41-42.

58. Industrial Commission, *Report,* 1: pt. 2, p. 303.

59. *New York Tribune,* December 22, 1898; Industrial Commission, *Report,* 1: pt. 2, pp. 303-07.

60. Industrial Commission, *Report,* 13:665.

61. Foraker, *Notes of a Busy Life,* 2: 331-32. William Randolph Hearst forced this explanation by revealing Foraker's ties with Standard Oil in a speech at Columbus, Ohio, on September 17, 1908. Hearst subsequently published several letters between Archbold and Foraker.

62. United States Senate, Committee on Privileges and Elections, *Campaign Contributions Testimony,* 62d Cong. 3d sess. (1913), 1: 792-93.

63. New York *World,* September 19, 1908. Foraker confirmed that he had met with Monnett, although he omitted the details of their discussion. Foraker, *Notes of a Busy Life,* 2: 333. Foraker defended his actions in testimony before the Senate Committee on Privileges and Elections in 1912, but he avoided comment on Monnett's charges. *Campaign Contributions Testimony.,* 2: 1275-1340.

64. Industrial Commission, *Report,* 1: pt 2, p. 321.

65. Ibid.

66. *Standard Oil Trust Cases, Record,* Malcolm Jennings testimony, pp. 312-23.

67. *Weekly Law Bulletin,* 41 (May 15, 1899): 301.

68. Industrial Commission, *Report,* 13: 600.

69. Ibid., p. 601.

70. New York *World,* March 6, 1899.

71. Ibid. March 7, 1899. Commentary from the Ohio press is collected in Industrial Commission, *Report,* 13: 661-62.

72. New York *World,* March 19, 1899.

73. Ibid. March 21, 1899.

74. Sandusky *Register,* April 17, 1899; see Industrial Commission, *Report,* 13:662 for other comments.

75. Attorney General of Ohio, *Annual Report* (Columbus, 1899), p. 42.

76. Industrial Commission, *Report,* 13:664.

77. Ibid.

78. Kline to Elliot, March 3, April 14, May 14, 1899, Elliot to J. O. Troup, February 6, March 9, 1899: cited by Hidy and Hidy, *Pioneering in Big Business,* p. 801, n. 45. On March 7, 1899, Monnett told the Cleveland *Plain Dealer* that if he had revealed Squire's name, "the trust would be after him at once." Elliot was astounded at this statement, privately declaring, "If we hired a man to bribe the Attorney General we ought to be as

familiar with his name as the Attorney General." Elliot to Troup, March 9, 1899, in Hidy and Hidy, *Pioneering in Big Business.*

79. James C. Bonbright and Gardiner C. Means give the following definition of a holding company: "A holding company may be defined in the broadest sense as any company having share capital which owns securities of one or more other companies. In a more restricted but more usual sense the definition is made to turn not on ownership in but on control over another company. A holding company may thus be defined in terms of its distinguishing characteristic as any company with share capital which is in a position to control or materially to influence the management of one or more other companies by virtue, in part at least, of its ownership of securities of the latter. A holding company may be classed as a pure holding company if its assets are composed almost entirely of the securities of other companies, and as a parent holding company (or parent company) if in addition to the ownership of such securities it conducts an operating enterprise as a directly owned property." *Encyclopedia of the Social Sciences* (New York: Macmillan Company, 1937), 9:403-04. Standard Oil of New Jersey was a parent company.

80. Hostile editorials appeared in the New York *World,* May 1, 1897, and in the *New York Herald,* May 13, 1897.

81. *United States* v. *E.C. Knight Company,* 156 U.S. 1 (1895).

82. James B. Dill, *The Statute and Case Law of New Jersey relating to Business Companies* (Camden, N.J.: Press of S. Chew and Sons Company, 1910), p. 80. Also see Dill's testimony before the Industrial Commission on New Jersey corporation laws in *Report,* 1: pt. 2, pp. 1077-1087, and Hidy and Hidy, *Pioneering in Big Business,* p. 308.

83. Bonbright and Means, "Holding Companies," p. 405.

84. *Brief for the United States,* no. 775, 1:73-79; Nevins, *Rockefeller,* 2:356; Hidy and Hidy, *Pioneering in Big Business,* p. 310.

85. On Burkett, see *National Cyclopaedia of American Biography* (New York: James T. White and Company, 1898-1951), 9:550; Marshall, *History of Courts and Lawyers of Ohio,* 1:259; Edgar B. Kinkead, "A Sketch of the Supreme Court of Ohio," *The Green Bag,* 7 (March 1895): 294. On Shauck, see *Dictionary of American Biography,* 17: 34-35. Also see *National Cyclopaedia,* 12:134; Marshall, *History of Courts and Lawyers of Ohio,* 1:259; and Kinkead, "Sketch of the Supreme Court of Ohio," p. 294. On Davis, see *National Cyclopaedia,* 9:551.

86. New York *World,* September 19, 1908.

87. D.J. O'Day to Elliot, April 26, May 3, 20, 1901, in Hidy and Hidy, *Pioneering in Big Business,* p. 802, n. 75.

88. Senate Committee on Privileges and Elections, *Campaign Contributions Testimony,* 1:797. Archbold sent a nearly identical letter to Mark Hanna. Ibid.

89. A letter from Archbold to Foraker, February 27, 1900, indicates the close ties between Sheets, Foraker, and the Standard organization: "Attorney-General Sheets has written a letter to Mr. Kline, in which he asks to have a time fixed for the oral argument of the contempt case. If this argument is to be simply a formal matter, we have no objection to it; otherwise, it might be well to have it postponed as long as possible, especially until after the next National and State conventions." J. E. —, "The History of the Standard Oil Letters," *Hearst's Magazine,* 21 (May 1912): 2211.

90. *Ohio ex rel. Monnett v. Buckeye Pipe Line Company, etc.,* 61 Ohio 520-23 (1900).

91. Ibid., at 548.

92. *U.S. v. Trans-Missouri Freight Association,* 166 U.S. 290 (1897); *U.S. v. Joint Traffic Association,* 171 U.S. 505 (1898); *U.S. v. Addystone Pipe and Steel Company,* 54 U.S. App. 723 (1898).

93. *Weekly Law Bulletin,* 44 (February 5, 1900): 51.

94. New York *World,* January 31, 1900. The more restrained New York *Tribune,* January 31, 1900, took a more realistic view. Its headlines simply declared: "Ohio Anti-Trust Law Upheld."

95. *Standard Oil Trust Cases, Record,* Henry Roeser testimony, pp. 258-65. The Bureau of Corporations found that Standard pipelines handled 90 to 95 percent of all Ohio crude in 1904. *Report on the Petroleum Industry* (Washington: Government Printing Office, 1907), 2:71.

96. Ohio Stat., 1898, p. 143, sec. 1.

97. *Weekly Law Bulletin,* 44 (December 31, 1900): 357.

98. Ibid. (December 24, 1900): 335.

99. The record stated: "This day this cause came to be heard upon the information against said defendant for contempt filed herein, and the evidence produced by said parties: On consideration whereof, the Court being fully advised in the premises does find that said defendant is not guilty. It is therefore considered and adjudged that said information be and it is dismissed, and that said defendant recover its costs herein expended." *United States v. Standard Oil Company of New Jersey,* 221 U.S. 1 (1911), *Record* (Washington: Government Printing Office, 1908), 22:525-26.

100. Quoted in *Literary Digest,* 21 (December 29, 1900): 796.

101. *New York Times,* January 27, 1906.

102. *Ohio ex rel. Wachenheimer v. Standard Oil Company of Oil, et al.,* 15 Ohio Circuit court Rep. (New Series) 212 (1907).

103. *New York Times,* July 11, 21, October 10, 11, 18, 19, 20, November 15, December 25, 1906; *The Outlook* 84 (October 27, 1906): 437-38, and (November 3, 1906): 50-551.

104. *New York Times,* November 15, 1906; *The Outlook,* 84 (November 24, 1906): 684; *Independent* 61 (November 22, 1906): 1192.

105. Apparently Ellis did not file suit against Ohio Standard because of the criminal charges pending against Rockefeller and other Ohio Standard leaders in Hancock County. Ohio, Attorney General's Office, *Annual Report, 1906-1907* (Columbus: F. J. Heer, 1907), p. xv.

106. Wade H. Ellis, *Proceedings of the National Conference on Trusts and Combinations* (New York: National Civic Federation, 1908), pp. 41-57; Wade H. Ellis, "The History of the Standard Oil Company in Ohio," *Ohio Magazine,* 4 (January 1908), pp. 1-10. Ohio, Attorney General's Office, *Annual Report, 1906-1907,* p. xv.

107. Ohio, Attorney General's Office, *Annual Report, 1910-1911* (Columbus: F. J. Heer, 1911), p. 26.

108. Carrington T. Marshall, *History of Courts and Lawyers of Ohio,* 4: 35-36.

109. Ohio, Attorney General's Office, *Annual Report, 1909-1910* (Springfield, Ohio: Springfield Publishing Company, 1910), pp. 26-27.

110. Ibid., p. 27; Ohio, Attorney General's Office, *Annual Report, 1910-1911,* p. 26.

111. Ohio, Attorney General's Office, *Annual Report, 1911-1912* (Springfield, Ohio: Springfield Publishing Company, 1912), p. xvii.

CHAPTER 2

1. Henry Clay Pierce to T. M. Campbell, October 15, 1907, quoted in Frederick U. Adams, *The Waters Pierce Case in Texas, Compiled from the Series of Press Articles Entitled: Battling with a Great Corporation* (St. Louis: Skinner and Kennedy, 1908), pp. 56-58; also see ibid., at 9-13; Allan Nevins, *Study in Power* (New York: Charles Scribner's Sons, 1953), 2:41-42.

2. *United States* v. *Standard Oil Company of New Jersey,* 173 Fed. 177 (1909), *Record* (Washington: Government Printing Office, 1908), Pierce testimony, 3:1073-74; Adams, *Waters Pierce Case,* pp. 11-13.

3. *U.S.* v. *Standard Oil, Record,* Pierce testimony, 3:1066-69; *Missouri* v. *Standard Oil (Indiana) Waters-Pierce Oil Company, and*

Republic Oil Company, 218 Mo. 1 (1909), *History of the Case* (Jefferson City, Mo.: H. Stephens, 1907), pp. 25-26; Adams, *Waters-Pierce Case,* p. 14.

4. *U.S.* v. *Standard Oil, Record,* Pierce testimony, 3:1066-67; *Missouri* v. *Standard Oil (Indiana), History,* pp. 26-27; Ralph Hidy and Muriel Hidy, *Pioneering in Big Business* (New York: Harper and Brothers, 1955), pp. 64, 72.

5. Ohio Senate, *Trust Investigation* (Columbus, 1899), pp. 370-71; Ida Tarbell, *The History of the Standard Oil Company* (New York: McClure, Phillips & Company, 1904), 2:46-48.

6. House of Representatives, Committee on Manufactures, *Investigation of Certain Trusts* (Washington: Government Printing Office, 1889), p. 573.

7. *Waters-Pierce Oil Company* v. *Texas,* 19 Tex. Civ. App. 1, 5-9 (1898); Nevins, *Study in Power,* 2:43.

8. J. D. Forrest, "Anti-Monopoly Legislation in the United States," *American Journal of Sociology,* 1 (January 1896): 413.

9. Dudley G. Wooten, in *Chicago Conference on Trusts* (Chicago: Civic Federation of Chicago, 1900), p. 43.

10. Robert W. Cotner, *James Stephen Hogg* (Austin: University of Texas Press, 1959), pp. 160-65; Texas, *House Journal* (Austin: Von Boeckman-Jones Company, 1889), p. 412; Texas, *Senate Journal* (Austin: Von Boeckman-Jones Company, 1889), p. 581; Tom Finty, Jr., *Anti-Trust Legislation in Texas* (Dallas: A.H. Belo Company, 1916), pp. 15-17.

11. Texas, *Statutes* (1889), p. 141.

12. George Rice to J.S. Hogg, May 7, 1889, quoted in Cotner, *Hogg,* p. 167.

13. Alwyn Barr, *Reconstruction to Reform: Texas Politics, 1876-1906* (Austin: University of Texas Press, 1971), p. 120.

14. *Hathaway* v. *State,* 36 Tex. Crim. Rep. 261 (1896), pp. 266-72. The Texas court system created by the judiciary act of 1891 was complex. The supreme court had three members and was the highest tribunal for civil cases. The court of criminal appeals also had three members and served as the supreme tribunal for criminal matters. Below the supreme court were five courts of civil appeals, each composed of three judges. There were more than one hundred district courts (the number varied over the years). The district court was the chief trial court for the state; a single judge presided over each. All judicial offices were elective. Frank M. Stewart and Joseph L. Clark, *The Constitution and Government of Texas,* rev. ed. (Boston: D. C. Heath and Company, 1933), pp. 111-28. Understandably,

this system provoked considerable criticism. Section 9 of the Populist party platform of 1898 declared: "We denounce the cumbrous judicial system of this State, consisting as it does of seven appellate courts, whose decisions conflict, entailing the State the expense of seven appellate courts without the virtue of one." Ernest R. Winkler, *Platforms of Political Parties in Texas* (Austin: University of Texas Press, 1916), p. 399.

15. At that time the leading decisions were: *State* v. *Hall,* 115 N.C. 811 (1894) and *State* v. *Jackson,* 36. Fed. 258 (1888).

16. On Clark, see Walter Prescott Webb, ed., *The Handbook of Texas* (Austin: Texas State Historical Association, 1952), 1:354-55; Cotner, *Hogg,* pp. 177, 321. On Johnson, see *National Cyclopaedia of American Biography* (New York: James T. White and Company, 1898-1951), 6:126; John William Leonard, ed., *Who's Who in Jurisprudence* (Brooklyn: A. W. Steven Printing Company, 1925), p. 783; and Johnson's testimony in *Proceedings and Reports of the Bailey Investigation Committee* (Austin: Von Boeckman-Jones Company, 1907), pp. 84-85. The briefs noted below reveal Clark and Johnson's views on the Texas antitrust law. S. C. T. Dodd, "The Present Legal Status of Trusts," *Harvard Law Review* 7 (October 1893): 157-69. On the conservative legal climate of the 1890s, see Arnold Paul, *Conservative Crisis and the Rule of Law,* rev. ed. (New York: Harper and Row, 1969).

17. *Hathaway,* at 261-72; Adams, *Waters Pierce Case,* p. 22.

18. *Hathaway,* at 263-78.

19. *In re Grice,* 79 Fed. 627 (1897), pp. 628-31.

20. Ibid., at 640, 646.

21. *Baker* v. *Grice,* 169 U.S. 284, 294 (1898).

22. *Waters-Pierce Oil Company* v. *Texas,* 19 Tex. Civ. App. 1, 3-4 (1898).

23. Ibid., at 5-9.

24. Ibid., at 17.

25. *Waters-Pierce Oil Company* v. *Texas,* 177 U. S. 28, 41 (1900).

26. Ibid., at 43.

27. Ibid., at 46.

28. Quoted in *Literary Digest,* March 31, 1900, p. 386.

29. Ibid.

30. W. L. Crawford, *Crawford on Baileyism* (Dallas: Eclectic News Bureau, 1907), p. 4; *Bailey Investigation Committee,* Francis testimony, p. 669.

31. Ibid.

32. For accounts of Bailey's career, see Sam Hanna Acheson, *Joe Bailey, The Last Democrat* (New York: Macmillan, 1932), and Bob

Charles Holcomb, "Senator Joe Bailey, Two Decades of Controversy" (Ph.D. diss., Texas Technological College, 1968). Bailey was a highly controversial figure during the first decade of the century, and several political tracts appeared concerning his relations with the Waters-Pierce Oil Company. William A. Cocke, *The Bailey Controversy in Texas* (San Antonio, Texas: The Cocke Company, 1908); E. G. Senter, *The Bailey Case Boiled Down* (Dallas: Flag Publishing Company, 1908); and Crawford, *Crawford on Baileyism,* all are critical of Bailey. Adams, *Waters Pierce Case,* primarily a defense of Henry Clay Pierce, has a more favorable attitude toward the Texas congressman. These accounts all suffer from intense partisanship, as well as a lack of knowledge of the then obscure workings of the trust. Since 1908 only two works have treated the Waters-Pierce affair. Acheson's biography, *Joe Bailey,* is an uncritical defense of Bailey's role in the controversy. Holcomb's dissertation, "Senator Joe Bailey," p. 219, finds Bailey guilty of "a definite breach of ethics."

33. *Bailey Investigation Committee,* Gibbs testimony, pp. 1025-56; Francis testimony, pp. 666-70; Bailey testimony, pp. 806, 809-12, 826-33.

34. Ibid., Bailey testimony, p. 838.

35. Ibid.

36. Ibid., p. 839.

37. Ibid. Bailey received an additional $1,500 loan from Pierce on June 13, 1900. Ibid., pp. 847-48.

38. On Sayers, see Joseph D. Sayers, "Anti-Trust Legislation," *North American Review* 169 (August 1898): 217; Barr, *Reconstruction to Reform,* pp. 214-16. Smith's speech is printed in *Chicago Conference on Trusts,* pp. 567-68.

39. Adams, *Waters Pierce Case,* pp. 29-30.

40. *Bailey Investigation Committee,* Bailey testimony, pp. 807, 841; Hardy testimony, p. 374; Rupert N. Richardson, *Colonel Edward M. House: The Texas Years, 1858-1912* (Abilene: Harden-Simmons University Publications in History, 1964), p. 168.

41. *Bailey Investigation Committee,* Bailey testimony, pp. 807, 843; T. S. Smith testimony, pp. 1027-28.

42. *U.S.* v. *Standard Oil, Record,* Pierce testimony, 3:1067; W. H. Gray, *The Rule of Reason in Texas* (Houston: W. H. Gray, 1912), p. 15.

43. *U.S.* v. *Standard Oil, Record,* Pierce testimony, 3:1080; Pratt testimony, 1:73-74.

44. Ibid., McNall testimony, 2:662-64; Gray, *Rule of Reason,* p. 15.

45. *Bailey Investigation Committee,* p. 1040.

46. Ibid. All the relevant documents are reprinted on pp. 1029-44.

47. *Chicago Conference on Trusts,* pp. 567-68.

48. *Bailey Investigation Committee*, p. 1040.

49. Acheson, *Bailey*, pp. 138-39; Holcomb, "Bailey," pp. 171-89.

50. Robert C. Cotner, ed., *Addresses and State Papers of James S. Hogg* (Austin: University of Texas Press, 1951), p. 479.

51. Ibid.

52. Ibid., p. 86.

53. *Dallas Morning News*, August 9, 1900.

54. Ibid.

55. Cotner, *Hogg*, p. 480; Holcomb, "Bailey," pp. 182, 236.

56. J. S. Hogg to Richard Coke, April 27, 1896, quoted in Cotner, *Hogg*, p. 476; also see pp. 410-11.

57. C. Vann Woodward, *Origins of the New South* (Baton Rouge: Louisiana State University Press, 1951), p. 370.

58. Texas, *House Journal* (1901), p. 47; Adams, *Waters Pierce Case*, p. 40.

59. *Bailey Investigation Committee*, J. R. Smith testimony, pp. 624-27; Hill testimony, pp. 660-61; Bailey testimony, p. 865.

60. Ibid., J. R. Smith testimony, pp. 1009-14; Junkin testimony, pp. 1014-17.

61. Ibid., Bailey testimony, p. 806; Thomas testimony, pp. 1018-25; Gibbs testimony, pp. 1025-27; T. S. Smith testimony, pp. 1027-28, 1042-44.

62. Ibid., p. 1004; Texas, *House Journal* (1901), pp. 182-84, 190-91.

63. Crawford, *Crawford on Baileyism*, p. 10.

64. *Bailey Investigation Committee*, Bailey testimony, pp. 850, 871; Holcomb, "Bailey," pp. 255-57, 259.

65. Carl Coke Rister, *Oil! Titan of the Southwest* (Norman: University of Oklahoma Press, 1949), pp. 50-65, is a complete account of the Spindletop strike. See Ralph Andreano, "The Emergence of New Competition in the American Petroleum Industry Before 1911" (Ph.D. diss., Northwestern University, 1960), p. 57, for the production figures.

66. Hidy and Hidy, *Pioneering in Big Business*, p. 394.

67. *New York Times*, August 2, 7, 1901; Finty, *Anti-Trust Legislation in Texas*, pp. 21-22, 64.

68. *State ex rel. Attorney General* v. *Waters-Pierce Oil Company*, 67 S.W. 1057 (1902). The Texas court followed the United States Supreme Court decision in *Connelly* v. *Union Sewer Pipe Company*, 184 U.S. 540 (1902), which overturned the Illinois Antitrust Act for granting agricultural producers a blanket exemption. Finty, *Antitrust Legislation in Texas*, pp. 18-19, 21-23; Adams, *Waters Pierce Case*, p. 49.

69. "The Foremost Democrat in Washington," *Current Literature* 40 (May 1906): 487. On Bailey's role in the passage of the Hepburn Act and the subsequent reaction, see Holcomb, "Bailey," pp. 302-50, 355-56.

70. David Graham Phillips, "The Treason of the Senate," *Cosmopolitan Magazine* 41 (July 1906): 267-76.

71. Holcomb, "Bailey," pp. 368-70; Rupert N. Richardson, *Texas: The Lone Star State* (New York: Prentice-Hall, 1943), pp. 370-71; James T. DeShields, *They Sat in High Places* (San Antonio: Naylor Company, 1940), pp. 395-96.

72. *Bailey Investigating Committee,* Johnson testimony, pp. 120-21, and Gruet testimony, pp. 201, 204; Adams, *Waters Pierce Case,* pp. 42-45; Hidy and Hidy, *Pioneering in Big Business,* pp. 449-50.

73. L. S. Flateau to Bailey, June 26, 1906, in *Bailey Investigating Committee,* p. 999; Gruet testimony, pp. 205, 213-18; Bailey testimony, pp. 863-64.

74. Ibid., Gruet testimony, pp. 217-18, 253; Adams, *Waters Pierce Case,* pp. 44-45.

75. Davidson knew that Bailey had been receiving reports about the attorney general's possession of information connecting the senator with the Waters-Pierce Oil Company. On June 21, 1906, Bailey wrote to Davidson denying the validity of any information connecting him with that company. Davidson made no attempt to contact Bailey about the matter. If Davidson had responded to Bailey's letter, he would have found out about the blackmail attempt. *Bailey Investigating Committee* reprints Bailey's letter to Davidson at p. 862.

76. *U.S.* v. *Standard Oil, Record,* Pierce testimony, 3:1065-90; Holcomb, "Bailey," pp. 374-75; Acheson, *Bailey,* p. 224.

77. Dallas *Morning News,* November 27, 1906; Texas, Attorney General's Office, *Report and Opinions of the Attorney General, 1906-1908* (Austin: Von Boeckman-Jones, 1909), p. 21.

78. Dallas *Morning News,* November 27, 1906; Cocke, *Bailey Controversy,* vol. 1, reprints the full Davidson-Bailey correspondence of fall 1906. Davidson's notice to produce documents is reprinted at p. 183.

79. Cocke, *Bailey Controversy,* pp. 185-86.

80. Ibid., pp. 187-97.

81. Quoted in Sam H. Acheson, *35,000 Days in Texas* (New York: Macmillan, 1938), p. 235.

82. Holcomb, "Bailey," p. 390.

83. Texas, *House Journal,* 30 Leg., Reg. sess. (Austin: Von Boeckman-Jones Company, 1907), pp. 108-51, 156; Texas, *State Journal,* 30 Leg.,

Reg. sess. (Austin: Von Boeckman-Jones Company, 1907), pp. 30-35, 64.

84. Texas, *House Journal,* 30 Leg., Reg. sess, p. 197; Texas, *Senate Journal,* 30 Leg., Reg. sess., p. 113.

85. *Bailey Investigating Committee,* on alterations: Gruet testimony, pp. 231-39; Naudin testimony, pp. 192-93; Hutchinson testimony, pp. 763-69, 796; Gruet, Jr., testimony, p. 291. On alcoholism: Johnson testimony, p. 153. On blackmail attempt: Gruet testimony, pp. 214-16; Flateau to Bailey, June 26, 1906, p. 999. On fee: Gruet testimony, p. 253.

86. Ibid., On Kirby Lumber: Kirby testimony, pp. 425-35. On Tennessee Central: Bailey testimony, p. 946. On Security Oil: Bailey testimony, p. 895. On Bailey's fees from Tennessee Central, also see Holcomb, "Bailey," pp. 421-23.

87. *Bailey Investigating Committee,* p. 1002; majority report, pp. 1059-73, 1099.

88. Texas, Attorney General's Office, *Report and Opinions, 1906-1909,* pp. 21-22; Gray, *Rule of Reason,* pp. 37-38.

89. *Waters-Pierce Oil Company* v. *Texas,* 48 Texas Civ. App. 162 (1907); Texas, Attorney General's Office, *Report and Opinions, 1906-1908,* p. 22; *Waters-Pierce Oil Company* v. *Texas,* 212 U.S. 86 (1909).

90. Davidson lost the Democratic nomination in 1910 to Oscar B. Colquitt. Prohibition was at least as important an issue in the campaign as either Bailey or the Waters-Pierce case. Colquitt supported prohibition, while Davidson opposed it. Holcomb, "Bailey," pp. 478-81; Acheson, *Bailey,* pp. 276-80; Richardson, *Texas,* p. 379.

91. Waters-Pierce profits: Crawford, *Crawford on Baileyism,* p. 10; *SONJ* v. *U.S.,* 221 U.S. 1 (1911), *Brief for the United States,* no. 725, 2:24. Oil trust earnings: Gray, *Rule of Reason,* p. 70.

92. Quoted in *Literary Digest,* 34 (June 15, 1907): 943.

93. Gray, *Rule of Reason,* pp. 16-21; Hidy and Hidy, *Pioneering in Big Business,* pp. 276, 393.

94. Andreano, "Emergence of New Competition," pp. 126, 133, 141; Gray, *Rule of Reason,* p. 26.

95. Cocke, *Bailey Controversy,* p. 883.

96. Texas, Attorney General's Office, *Report and Opinions, 1906-1908,* pp. 24-25; Gray, *Rule of Reason,* p. 40.

97. Finty, *Anti-Trust Legislation in Texas,* p. 65; Gray, *Rule of Reason,* p. 45.

98. Gray, *Rule of Reason,* p. 48; *Dallas Morning News,* December 8, 1909.

99. Gray, *Rule of Reason,* p. 56; *New York Times,* July 22, 1913; Isaac F. Marcosson, *Black Golconda* (New York: Harper and Brothers, 1924), p. 233.

100. Gray, *Rule of Reason,* p. 49; Finty, *Anti-Trust Legislation in Texas,* p. 21; *Dallas Morning News,* December 8, 1909.

101. In addition to the six suits covered in this chapter, a county attorney general filed an unsuccessful suit for penalties against Waters-Pierce in 1895. For details see *Bailey Investigating Committee,* pp. 114, 330-31, 393-400, 704-09.

102. Ibid., p. 808.

103. For example, see Carl Solberg, *Oil Power: The Rise and Imminent Fall of an American Empire* (New York: New American Library, 1976), pp. 53-68; Anthony Sampson, *The Seven Sisters* (New York: Bantam Books, 1975), pp. 44-47.

CHAPTER 3

1. Henry D. Lloyd, *Wealth Against Commonwealth* (Washington: National Home Library Association, 1936). Contrasting evaluations of Lloyd's accuracy are found in Allan Nevins, *Study in Power* (New York: Charles Scribner's Sons, 1953), 2:330-31; and Chester M. Destler, *Henry Demarest Lloyd and the Empire of Reform* (Philadelphia: University of Philadelphia Press, 1963), p. 300-05. Tarbell's series was reissued in 1904 by McClure, Phillips and Company as *The History of the Standard Oil Company.* Isobel C. Sheiffer, "Ida M. Tarbell and Morality in Big Business" (Ph.D diss., New York University, 1967), is the fullest account of Tarbell's career. President Roosevelt's treatment of Standard Oil is examined below in chapter 5. Peter Collier and David Horowitz, *The Rockefellers* (New York: New American Library, 1976), p. 44, discuss John D. Rockefeller's personal reaction to the negative publicity.

2. New York State Assembly, *Proceedings of the Special Committee Appointed to Investigate Abuses in the Management of Railroads* (New York: Evening Post Steam Presses, 1879-1880); New York Senate, *Report of the Committee on General Laws on the Investigation Relative to Trusts* (Troy, N.Y.: Troy Press Company, 1888); United States House of Representatives, Committee on Manufactures, *Investigation of Certain Trusts* (Washington: Government Printing Office, 1889); United States Industrial Commission, *Report on Trusts and Industrial Combinations* (Washington: Government Printing Office, 1900); United States Bureau of Corporations, *Report on the Transportation of Petroleum* (Washington: Government Printing Office, 1906); United States Bureau of Corporations, *Report on the Petroleum Industry* (Washington: Government Printing Office, 1907); United States Interstate Commerce Commission, *Report of the*

Investigation of Railroad Discriminations and Monopolies of Oil, House Doc. 606, 59th Cong., 2d sess. (Washington: Government Printing Office, 1907).

3. Production statistics are from Ralph Andreano, "The Emergence of New Competition in the American Petroleum Industry before 1911" (Ph.D. diss., Northwestern University, 1960), p. 57.

4. Ibid., pp. 24, 63.

5. United States Industrial Commission, *Report on Trusts,* 2:222.

6. Ibid., pp. 222-27, reprints of the 1891 and 1897 laws.

7. Quoted in *Standard Oil Company, et al.* v. *State,* 117 Tenn. 618 (1907), p. 627.

8. Ibid., at 627-29.

9. Ibid., p. 632; Ralph Hidy and Muriel Hidy, *Pioneering in Big Business* (New York: Harper and Brothers, 1955), p. 196; see above chapter 2 for Pierce's early battles with Chess, Carly & Company.

10. *Standard Oil* v. *State,* 117 Tenn. 618, 632-33 (1907).

11. Ibid., at 634.

12. Ibid.

13. Ibid, at 634-35.

14. Ibid., at 635.

15. Ibid., at 636.

16. Ibid., at 629; *New York Times,* February 4, 1904.

17. *Standard Oil* v. *State,* at 638-39.

18. *Tennessee ex rel. Cates* v. *Standard Oil Company of Kentucky,* 120 Tenn. 86 (1908), pp. 92-93.

19. Ibid., at 172.

20. Ibid., at 158.

21. *Standard Oil Company of Kentucky* v. *Tennessee,* 217 U.S. 413, 420 (1910), p. 420.

22. *Literary Digest* 40 (May 21, 1900): 1017.

23. Ibid.

24. Hidy and Hidy, *Pioneering in Big Business,* pp. 401, 460, 475.

25. Ibid., p. 275; Carl Coke Rister, *Oil! Titan of the Southwest* (Norman: University of Oklahoma Press, 1949), pp. 31-34.

26. Hidy and Hidy, *Pioneering in Big Business,* p. 276.

27. Kansas City *Star,* April 30, 1896, and January 18, 1903, quoted in Rister, *Oil!,* pp. 35-36.

28. Hidy and Hidy, *Pioneering in Big Business,* pp. 394-95.

29. Report of L. W. Keplinger, Commissioner of the Supreme Court, Kansas Attorney General's Office, *Eighteenth Biennial Report of the Attorney General of Kansas, 1911-12* (Topeka: State Printing Office, 1912), pp. 66-68.

30. Isaac F. Marcosson, "The Kansas Oil Fight," *World's Work* 10 (May 1905): 6159; Charles M. Harger, "Kansas' Battle for Its Oil Interests," *American Monthly Review of Reviews,* 31 (April 1905): 472.

31. Richard L. Douglas, "A History of Manufactures in the Kansas District," *Collections of the Kansas State Historical Society,* 11 (Topeka: State Printing Office, 1910), p. 193.

32. Ida M. Tarbell, "Kansas and the Standard Oil Company," *McClure's Magazine* 25 (September-October 1905), pp. 472-72; Marcosson, "Kansas Oil Fight," pp. 6159, 6166; Harger, "Kansas' Battle," p. 473; Harold F. Williamson et al., *The American Petroleum Industry: The Age of Energy* (Evanston: Northwestern University Press, 1963), p. 91.

33. *United States* v. *Standard Oil Company of New Jersey,* 173 Fed. 177 (1909), *Record,* Knapp testimony, 3:1233-36; Robertson testimony, p. 1183; Koontz testimony, p. 1249; Leland testimony, 15:2217-18; Bureau of Corporations, *Report of the Commissioner of Corporations on the Transportation of Petroleum* (Washington: Government Printing Office, 1906), pp. 370-94; Tarbell, "Kansas and the Standard Oil Company," pp. 473-74.

34. F. S. Barde, "The Oil Fields and the Pipe Lines of Kansas," *The Outlook* 80 (May 6, 1905): 23; Tarbell, "Kansas and the Standard Oil Company," p. 479; Hidy and Hidy, *Pioneering in Big Business,* p. 672.

35. William E. Connelley, "The Kansas Oil Producers Against the Standard Oil Company," *Collections of the Kansas State Historical Society,* 9 (Topeka: State Printing Office, 1906), p. 94; Tarbell, "Kansas and the Standard Oil Company," p. 610.

36. Marcosson, "Kansas Oil Fight," p. 6161.

37. Robert Clark, "Breaking Up a State Machine," *Cosmopolitan* 37 (October 1904): 665-70; William F. Zornow, *Kansas: A History of the Jayhawk State* (Norman: University of Oklahoma Press, 1957), pp. 209-11.

38. Kansas, *Senate Journal,* 14th biennial sess., 1905 (Topeka: State Printing Office, 1905), p. 16; Kansas, *House Journal,* 14th biennial sess., 1905 (Topeka: State Printing Office, 1905), pp. 18-19.

39. Kansas, *Senate Journal,* 1905, p. 38; Marcosson, "Kansas Oil Fight," p. 6161.

40. Connelley, "Kansas Oil Producers," pp. 94-101; Marcosson, "Kansas Oil Fight," pp. 6161-62.

41. Hidy and Hidy, *Pioneering in Big Business,* p. 675.

42. Tarbell, "Kansas and the Standard Oil Company," p. 616.

43. Marcosson, "Kansas Oil Fight," pp. 6161-62.

44. Harger, "Kansas' Battle," p. 473; *The Outlook* 79 (February 25, 1905): pp. 463-64; Barde, "Oil Fields and Pipe Lines," pp. 29-30.

45. John J. McLaurin, "The Oil Situation in Kansas," *The Outlook* 80 (June 17, 1905): 429.

46. Charts tracing the passage of these bills are given in Kansas, *Senate Journal,* 1905, pp. 752, 782; Kansas, *House Journal,* 1905, pp. 1211, 1227. Also see Connelley, "Kansas Oil Producers," pp. 99-100; Harger, "Kansas' Battle," p. 471.

47. Kansas, Attorney General's Office, *Fifteenth Biennial Report of the Attorney General of Kansas, 1905-1906* (Topeka: State Printing Office, 1906), pp. 26-27.

48. In early 1905, state Senator Porter explained: "My plan was to create legislative competition and then other competition would follow. I wanted to encourage independent refineries in the State so that all the oil would not have to go to one corporation." Quoted in Marcosson, "Kansas Oil Fight," p. 6161.

49. Kansas Stat. 1889, sec. 361.

50. *State* v. *Kelly,* 71 Kan. 811, 830 (1905).

51. United States Bureau of Corporations, *Report on the Petroleum Industry,* pt. 1. See chapter 6 below for an account of the report.

52. Kansas, Attorney General's Office, *Fifteenth Biennial Report, 1905-1906,* p. 27; *Emporia Gazette,* March 5, 1906; *Topeka State Journal,* March 5, 1906.

53. Kansas, Attorney General's Office, *Eighteenth Biennial Report, 1911-1912,* p. 63.

54. Ibid, pp. 59-63.

55. The feature articles are an important source of the Kansas episode and are referred to throughout the notes. See *Literary Digest* 30 (February 25, 1905): 271-72, and (March 4, 1905): 303-05, for newspaper reaction.

56. William Allen White, "Political Signs of Promise," *The Outlook* 80 (July 15, 1905): 669.

57. Tarbell, "Kansas and the Standard Oil Company," p. 622.

58. *The Independent* 63 (March 2, 1905): 453.

59. Quoted in *Literary Digest* 30 (February 25, 1905): 271.

60. Edward W. Hoch, "Kansas and the Standard Oil Company," *The Independent* 58 (March 2, 1905): 463.

61. Ibid.

62. Statistics on production and prices of mid-continent oil are given in Douglas, "History of Manufactures in the Kansas District," p. 193.

63. Ibid., p. 199. Also see *The Outlook* 81 (September 9, 1905): 55-56; 82 (February 10, 1906): 283; and 84 (April 14, 1906): 820-21, for continuing producer difficulties.

64. Keplinger report, in Kansas, Attorney General's Office, *Eighteenth Biennial Report, 1911-1912,* p. 91.

65. Ibid., p. 68.

66. Ibid., p. 91.

67. Ibid., pp. 69-81.

68. Ibid., p. 91.

69. Kansas Stat. 1889, sec. 359.

70. *Standard Oil Company of New Jersey* v. *United States,* 221 U.S. 1 (1911), *Brief for the United States,* no. 725, 1:82-84.

71. Kansas, Attorney General's Office, *Eighteenth Biennial Report, 1911-1912,* p. 92.

72. Ibid., p. 11.

73. Ibid., pp. 11-12.

74. Ibid., pp. 92-94.

75. William E. Connelley, *A Standard History of Kansas and Kansans* (Chicago, 1918), 3:1208. Also see John D. Bright el al., *Kansas: The First Century* (New York, 1956), 1:542-43, 9:557-58.

76. Zornow, *Kansas,* pp. 215-21.

77. Dana Gatlin, "What I Am Trying to Do?: An Interview with Hon. W. R. Stubbs," *World's Work* 24 (May 1912): 59-60. Also see *Dictionary of American Biography* (New York: Charles Scribner's Sons, 1944), 21:677-78, for a brief account of Stubbs's mixture of "private business efficiency and economy" and progressive politics.

78. Keplinger report, pp. 68-80.

CHAPTER 4

1. Louis G. Geiger, *Joseph W. Folk* (Columbus: University of Missouri Press, 1953), traces Folk's political career. The most complete treatment of Hadley is Lloyd E. Worner, "The Public Career of Herbert S. Hadley" (Ph.D. diss., University of Missouri, 1946).

2. Missouri Attorney General's Office, *Missouri ex inf. Herbert S. Hadley* v. *Standard Oil Company (of Indiana), Waters-Pierce Oil Company and Republic Oil Company: History of the Case, Issues Made by the Pleadings, Statement of Facts, Brief and Argument* (Jefferson City: H. Stephens Printing Company, 1907), pp. 22-92 (hereafter cited as *History of the Case).*

3. Ibid., pp. 25-28, 46-53; *United States* v. *Standard Oil Company of New Jersey,* 173 Fed. 177 (1909), *Brief for the United States,* 1:134-36; *Record,* Pierce testimony, 3:1065-90.

4. Indiana Standard replaced Standard Oil of Kentucky as the Oil Trust's marketing agent in northern Missouri.

5. *History of the Case,* pp. 28-29, 46-53, 67-70; *U.S.* v. *Standard Oil, Record,* L. C. Lohman testimony, 2:932-38, and John Burrows testimony, 3:1056-61.

6. *History of the Case,* pp. 30-31, 82-83; *U.S.* v. *Standard Oil,* 177 (1909), *Brief for U.S.,* 2:349-55.

7. *History of the Case,* pp. 71-79.

8. Ibid., p. 91.

9. Ibid., pp. 86-88; *U.S.* v. *Standard Oil, Brief for U.S.,* 2:352-53.

10. Sherman Morse, "The Taming of Rogers," *American Magazine* 62 (July 1906): 252; Frank C. Lockwood, "Governor Hadley of Missouri," *Independent* 66 (April 8, 1909): 744; Worner, "Herbert S. Hadley," pp. 93-96.

11. *Missouri ex inf. Hadley* v. *Standard Oil Company, et al.,* 218 Mo. 1, 36-43 (1909).

12. *History of the Case,* pp. 1-18.

13. "The Standard Oil Company at the Bar," *Arena* 35 (March 1906): 307. The New York *World* on January 5 and 6, 1906, and the *New York Times* on January 6, 1906, also detailed attempts to escape the subpoena servers.

14. *New York American,* January 16, 1906, quoted in *Arena* 35 (March 1906): 309.

15. Ibid. Hadley's agents even prevented John D. Rockefeller from visiting his newborn grandchild. The New York *World,* March 23, 1906, headlined: "Grandson Born to J. D. Rockefeller. And He, Mewed Up in His Lakewood Fort, Could Only Rejoice By Phone."

16. St. Louis *Post-Dispatch,* February 18, 1906, cited by Hazel Tutt Long, "Attorney General Herbert S. Hadley versus the Standard Oil Trust," *Missouri Historical Review* 35 (1940): 180.

17. Morse, "Taming of Rogers," p. 234; *New York Times,* January 6, 9, 1906.

18. Morse, "Taming of Rogers," p. 234; *History of the Case,* p. 10; *New York Times,* January 9, 1906. See summary of press reaction in *Literary Digest* 32 (January 20, 1906): 75-77; "Standard Oil Company at the Bar," p. 309.

19. *History of the Case,* p. 13; *Missouri ex inf. Hadley* v. *Standard Oil Company of Indiana, et al.,* 194 Mo. 124 (1906).

20. *New York Times,* February 1, 1906.

21. *Missouri ex inf. Hadley,* 194 Mo. 124 (1906); *New York Times,* February 27, 1906.

22. New York *World,* March 25, 1906.

23. *New York Times,* March 25, 1906; New York *World,* March 25, 1906.

24. New York *World,* March 25, 1906.

25. New York *World,* January 11, 1906.

26. *U.S.* v. *Standard Oil, Record,* Pierce testimony, 3:1065-90.

27. *Missouri ex. inf. Hadley,* 218 Mo. 1, 339 (1909).

28. For the impact of the investigation on Hadley's political career, see H. J. Haskell, "The People His Clients," *The Outlook* 88 (March 28, 1908): 719.

29. *Missouri ex. inf. Hadley,* 218 Mo. 1, 368-85 (1909).

30. Ibid., at 462.

31. Ibid., at 460.

32. Ibid., at 465-69.

33. Ibid, at 459-60.

34. Ibid., at 465-69.

35. Ibid., at 470-73.

36. Ibid., at 474.

37. Ibid., at 477.

38. Ibid., at 489.

39. Ibid., at 483.

40. *Literary Digest* 38 (February 13, 1909): 235.

41. *Missouri ex inf. Hadley,* 218 Mo. 1, 477 (1909).

42. *Standard Oil Company of Indiana* v. *Missouri ex inf. Hadley; Republic Oil Company* v. *Missouri ex inf. Hadley,* 224 U.S. 270 (1912).

43. *New York Times,* March 21, 1906.

44. *National Petroleum News* 4 (May 1912): 32; *Oil and Gas Journal* 10 (May 2, 1912): 3.

45. *National Petroleum News* 5 (March 1913): 16, 33.

46. *New York Times,* February 13, 1913; "Trust Busting vs. Regulation," *The Outlook* 104 (August 2, 1913): 731.

47. "Trust Busting vs. Regulations," pp. 731-32.

48. Governor Elliot W. Major, "Veto Message," in *Messages and Proclamations of the Governors of the State of Missouri,* ed. Sarah Guiter and Floyd C. Shoemaker (Columbia, Mo.: State Historical Society, 1928), p. 114.

49. Ibid., p. 115.

50. *New York Times,* May 11, 1913.

51. Ibid., May 17, 20, June 22, 1913. Refer to chapter 7 for the federal dissolution of Jersey Standard.

52. *Missouri ex inf. Hadley* v. *Standard Oil Company, et al.,* 251 Mo. 271, 272 (1913).

53. Justice Brown's dissent did not appear in the official state report, but it is given at 158 S.W. 601, 602-03 (1913).

54. Ibid., at 603.

55. Worner, "Herbert S. Hadley," pp. 122-23.

56. Nineteen of these suits reached the courts. Counting appeals, courts at various levels handed down a total of thirty-nine separate decisions. The United States Supreme Court ruled in these cases on four occasions.

57. On the Nebraska suit, see *New York Times*, November 7, 1901; Nebraska Attorney General's Office, *Report of the Attorney General, 1901-1902* (Lincoln: State Journal Company, 1902), p. 42. For Minnesota: *State ex rel. Edward T. Young* v. *Standard Oil Company*, 111 Minn. 85 (1910); Minnesota Attorney General's Office, *Biennial Report, 1912-1913* (Minneapolis: Syndicate Printing Company, 1913). Iowa: *State* v. *Standard Oil Company*, 150 Iowa 46 (1911); Iowa Attorney General's Office, *Biennial Report, 1909-1910* (Des Moines: Emory H. English, 1911). Arkansas: Ralph Hidy and Muriel Hidy, *Pioneering in Big Business* (New York: Harper and Brothers, 1955), pp. 683, 699, 772 n.2 notes the Arkansas suit and subsequent investigation. Oklahoma: Hidy and Hidy, *Pioneering in Big Business*, p. 683, notes the initiation of antitrust proceedings, and *The Oil and Gas Journal* 9 (July 14, 1910): 14-15, gives the details of the 1910 settlement. Mississippi: *Philadelphia Press*, July 24, 1907, p. 4; Mississippi Attorney General's Office, *Biennial Report, 1907-1909* (Nashville: Braden Printing Company, 1909), p. 21; *Biennial Report, 1910-1911*, pp. 10-11; *Biennial Report, 1912-1913*, pp. 12-13.

58. Standard Oil contributed $250,000 to the Republican presidential campaign fund in both 1896 and 1900. The total GOP fund was $3.5 million in 1896 and $2.5 million in 1900. Only one other contributor in 1900 equaled Standard's donation. Edward M. Sait, *American Parties and Elections* (New York: The Century Company, 1927), pp. 507-08. Standard Oil also made substantial contributions in this period to the campaign chest of Senator Hanna and Congressman C. H. Grosvenor of Ohio. J. E. ————, "The History of the Standard Oil Letters," *Hearst's Magazine* 21 (June 1912): 2365-68, 2376c.

59. Archbold to Foraker, December 18, 1902, quoted in Everett Walters, *Joseph Benson Foraker* (Columbus: Ohio History Press, 1948), p. 274.

60. *The Oil and Gas Journal* 9 (July 14, 1910): 14-15.

61. *New York Times*, August 29, 1900, November 7, 1901; Nebraska Attorney General's Office, *Report of the Attorney General, 1901-1902*, p. 42.

CHAPTER 5

1. On New Jersey Standard's international operations, see Ralph Hidy and Muriel Hidy, *Pioneering in Big Business* (New York: Harper and Brothers, 1955), chaps. 5, 17, 18, 19; Allan Nevins, *John D. Rockefeller: The Heroic Age of American Enterprise*, vol. 1 (New York: Charles Scribner's Sons, 1940), chap. 5. For brief sketches of the Standard Oil leaders by recent writers, see Peter Collier and David Horowitz, *The Rockefellers* (New York: Signet, 1976), pp. 43-46; Anthony Sampson, *The Seven Sisters* (New York: Viking Press, 1975), pp. 39-41, and Carl Solberg, *Oil Power* (New York: Mentor, 1976), pp. 42-51.

2. *United States* v. *Standard Oil Company of New Jersey*, 173 Fed. 177 (1909), *Record* (Washington: Government Printing Office, 1909), 7:petitioner's exhibits 1, 11, 31, 31A; also see summary in *Brief of Facts for Petitioner*, 1:80-91. *Independent* 63 (September 26, 1907): 715.

3. United States Bureau of Corporations, *Report on the Petroleum Industry, Part I* (Washington: Government Printing Office, 1907), pp. 7-9; Ralph Andreano, "The Emergence of New Competition in the American Petroleum Industry before 1911" (Ph.D. diss., Northwestern University, 1960), p. 57; George W. Stocking, *The Oil Industry and the Competitive System* (Boston: Houghton Mifflin, 1925), p. 50.

4. Bureau of Corporations, *Report on the Petroleum Industry, Part I*, pp. 9-13; Andreano, "Emergence of New Competition," p. 282; Arthur M. Johnson, *Petroleum Pipelines and Public Policy, 1906-1959* (Cambridge: Harvard University Press, 1967), p. 55.

5. Bureau of Corporations, *Report on the Petroleum Industry, Part I*, pp. 24-38.

6. Ibid., pp. 13-16; Nevins, *Heroic Age*, 2:569-70; Andreano, "Emergence of New Competition," p. 282.

7. *U.S.* v. *Standard Oil, Record*, 8:petitioner's exhibits 376, 377, 388, 389, 394; Stocking, *The Oil Industry and the Competitive System*, p. 52; Andreano, "Emergence of New Competition," p. 282.

8. *U.S.* v. *Standard Oil, Brief of Facts for Petitioner*, 2:427-639, for an extensive catalog of anticompetitive practices.

9. Ibid., 1:189-219; Bureau of Corporations, *Report on the Petroleum Industry, Part II: Prices and Profits* (Washington: Government Printing Office, 1907), p. 11-21, 321-440.

10. Ibid.; *Brief of Law for Petitioner*, pp. 30-31; *Brief of Facts for Petitioner*, 1:170-71.

11. G. A. Copeland to Miller, June 24, 1891, Justice Department File

8247-J.I. and Miller to Copeland, July 1, 1891, JD Misc. Bk. No. 5, p. 340, Record Group 60, National Archives, Washington, D.C.

12. Miller to Brinsmade, August 14, 1891, JD Instr. Bk. No. 14, p. 517.

13. G. D. Reynolds to Miller, August 20, 1891, JD File 8247-8463, and Miller to Reynolds, September 14, 1891, JD File 8247-8463.

14. S. R. Kepler to Miller, May 13, 1892, JD File 8247-5101; Raleigh *State Chronicle,* June 26, 1890.

15. Kepler to Miller, July 16, 1892, JD File 8247-7138.

16. Ibid., August 8, 1892, JD File 8247-7886.

17. Ibid., August 17, 1892, JD File 8247-8382; Miller to Kepler, August 20, 1892, JD File 8247-8382.

18. Independent Oil Men to Olney, June 7, 1894, JD File 8247-6604.

19. Rice to Griggs, November 5, 1898, JD File 8247-18000; Griggs to Rice, November 15, 1898, JD File 8247-18000.

20. Rice to Griggs, January 10, 1899, JD File 8247-576; ibid., February 16, 1899, JD File 8247-2730; Griggs to Rice, March 14, 1899, JD File 8247-2730.

21. Rice to Griggs, April 17, 1899, JD File 8247-6299.

22. Ibid., April 24, 1899, JD File 8247-6529; ibid., May 1, 1899, JD File 8247-6893; ibid., May 15, 1899, JD File 8247-90.

23. Rice to McKinley, May 5, 1899, JD File 8247-70.

24. A. G. Stafford to Griggs, May 31, June 19, 1899, JD File 8247-8564.

25. Griggs to Stafford, June 20, 1899, JD File 8247-8564.

26. Stafford to Griggs, July 26, 1899, JD File 8247-16085.

27. United States Industrial Commission, *Report of the Industrial Commission on Trusts and Industrial Combinations* (Washington, D.C.: Government Printing Office, 1900), 1:pt. 2, p. 671.

28. Ibid., at 728.

29. United States Attorney General, *Annual Report, 1891* (Washington: Government Printing Office, 1891), pp. 3-15, details the greatly increased workload. On the antitrust activities of the Justice Department in the 1890s, also see William Letwin, *Law and Economic Policy in America* (New York: Random House, 1965), chap. 4; Hans Thorelli, *Federal Antitrust Policy* (Baltimore: Johns Hopkins University Press, 1955), chap. 7; Homer Cummings and Carl McFarland, *Federal Justice* (New York: Macmillan Company, 1937), pp. 317-27; and James S. Easby-Smith, *The Department of Justice* (Washington, D.C.: W. H. Lowdermilk & Co., 1904), pp. 18-44.

30. United States House of Representatives, *Document No. 234,* 54th Cong., 1st sess., p. 234.

31. U. S. Attorney General, *Annual Report, 1891*, p. xxvii; Letwin, *Law and Economic Policy*, pp. 104-05.

32. Walton Hamilton, *The Pattern of Competition* (New York: Columbia University Press, 1940), pp. 59-62; Cummings and McFarland, *Federal Justice*, pp. 320-21.

33. Hamilton, *Pattern of Competition*, pp. 62-82.

34. *United States v. E. C. Knight Company*, 156 U.S. 1 (1895).

35. *United States v. Trans-Missouri Freight Association*, 166 U.S. 290 (1897). Chapter 6 deals with the question of reasonable restraints on interstate trade.

36. *United States v. Joint Traffic Association*, 171 U.S. 505 (1898).

37. *United States v. Addystone Pipe & Steel Company*, 175 U.S. 211 (1899).

38. Ibid., at 240.

39. Roosevelt to H. H. Kohlsaat, August 7, 1899, in Elting E. Morison, ed., *The Letters of Theodore Roosevelt* (Cambridge: Harvard University Press, 1951-1954), 2:1045.

40. Roosevelt to H. C. Lodge, August 10, 1899, in ibid., p. 1048.

41. Theodore Roosevelt, "Message to New York State Legislature, " in *The Works of Theodore Roosevelt* (New York: Charles Scribner's Sons, 1926), 15:43.

42. Ibid., pp. 44-45.

43. Ibid., p. 46.

44. Theodore Roosevelt, "Message at the Opening of the First Session of the 59th Congress, December 5, 1905," in William Griffith, ed., *The Roosevelt Policy* (New York: Current Literature Publishing Company, 1919), 1:326.

45. Roosevelt, "Message to New York State Legislature," in *Works of Roosevelt*, 15:45.

46. United States, *Statutes at Large*, 32:823; Cummings and McFarland, *Federal Justice*, p. 329.

47. Cummings and McFarland, *Federal Justice*, p. 329.

48. *Northern Securities Company v. United States*, 193 U.S. 197, 326 (1904).

49. Roosevelt to Knox, June 23, 1904, box 26, Knox Papers, Manuscript Division, Library of Congress, Washington, D.C.

50. Rice to Knox, June 26, 1901, JD File 8247-9693.

51. Ibid., August 9, 1901, JD File 8247-12443.

52. Ibid., August 16, 1901, JD File 8247-12443.

53. Ibid., January 6, 1902, JD File 8247-413.

54. Knox to Rice, January 9, 1902, JD File 8247-413.

55. Rice to Knox, January 10, 1902, JD File 8247-11256.

56. Ibid., August 16, 17, 1902, JD File 8247-14299.

57. Rice to Roosevelt, August 1, 1902, JD File 8247-12607.

58. Ibid., August 8, 1902, JD File 8247-12907; ibid., Roosevelt, August 22, 1902, JD File 8247-14618; ibid., August 29, 1902, JD File 8247-15087; ibid., September 12, 1902, JD File 8247-15562; ibid., October 28, 1902, JD File 8247-17685.

59. Ibid., December 30, 1902, JD File 8247-85.

60. H. J. Hammond to Knox, April 4, 1903, JD File 8247-5323; William H. Davis to M. D. Purdy (Assistant Attorney General), April 11, 1903, JD File 8247-5782; C. H. Buck to Knox, July 28, 1903, JD File 8247-11522; C. M. Gray to Roosevelt, February 24, 1904, JD File 46330-51758; Fred L. King to Roosevelt, November 30, 1904, JD File 46330.

61. *Rice v. Standard Oil Co.,* 134 Fed. 464 (1905).

62. Ibid., at 634.

63. Ibid., at 641.

64. New York *World,* March 2, 1905.

65. *Congressional Record,* 36, pt. 2, 57th Cong., 2d sess., p. 2005.

66. The *New York American,* February 12, 1903, was the first to make public the full text of the telegram.

67. John D. Rockefeller, Jr., papers, quoted in Nevins, *Heroic Age,* 2:516.

68. *New York Times,* February 8, 1903. Other papers that printed Roosevelt's version of the message of February 8, 1903, included the New York *World, New York American,* and *New York Tribune.*

69. United States, *Statutes at Large,* 32:825.

70. *New York Times,* February 10, 1903.

71. United States Senate, Committee on Privileges and Elections, *Campaign Contributions* (Washington: Government Printing Office, 1913), Archbold testimony, 2:123. See Henry F. Pringle, *Theodore Roosevelt* (New York: Harcourt, Brace, 1931), 1:357-58, for a summary of campaign contributors to Roosevelt in 1904.

72. New York *World,* October 1, 1904.

73. Taft to Helen H. Taft, August 22, 1912, series 2, Taft Papers, Manuscript Division, Library of Congress; Roosevelt to Cortelyou, October 26, 1904, in Morison, *Letters of Roosevelt,* 4:995-96.

74. Senate Committee on Privileges and Elections, *Campaign Contributions,* Archbold testimony, 1:129-31.

75. United States Bureau of Corporations, *Report on the Transportation of Petroleum* (Washington: Government Printing Office, 1906), p. xix. The conservative New York *Sun,* April 29, 1905, declared, "The report

[on the beef trust] recently submitted does not quite get at what the consuming public most wants to know." The *New York Press,* March 21, 1905, called the report "silly" and "disgraceful." For the president's reaction, see Roosevelt to Garfield, April 14, 1905, in Morison, *Letters of Roosevelt,* 9:1159-60.

76. Bureau of Corporations, *Report on the Transportation of Petroleum,* p. xix.

77. *Congressional Record* 40, pt. 7, 59th Cong. 1st sess., p. 6358.

78. "The President's Arraignment of Standard Oil," *Literary Digest* 32 (May 12, 1906): 716; "Standard Oil on the Rack," *Public Opinion* 40 (May 19, 1906): 613. JD File 46330 contains several complaints to Justice Department officials about Standard Oil during 1905 and 1906.

79. *New York Times,* June 23, 1906. On June 21, 1906, Taft suggested that the following statement be issued: "It is charged, in many complaints already filed in this Department, that the Standard Oil Company has maintained a burdensome monopoly of interstate commerce in oil and other products . . . by methods subjecting it to prosecution under both clauses of the so-called Anti-Trust Law. Investigation of the company's business and affairs will therefore continue for the purpose of restraining any other violations of law looking to maintenance of an oppressive monopoly, if found to exist. This will be done either by civil or criminal actions, as circumstances will prove to be most effective." Taft to Moody, June 21, 1906, vol. 14, no. 2648, Moody Papers, Manuscript Division, Library of Congress.

80. Frank B. Kellogg, "Speech of October 23, 1907," in *Proceedings of the National Conference on Trusts and Combinations* (New York: National Civic Federation, 1908), p. 217. For a sympathetic account of Kellogg's career, see David Bryn-Jones, *Frank B. Kellogg* (New York: G. P. Putnam's Sons, 1937).

81. *New York Times,* June 23, 1906.

82. Garfield, "1906 Diary," June 25, 26, 27, 29, September 22, 29, October 1, November 12, box 7, Garfield Papers, Manuscript Division, Library of Congress; ICC to Moody, July 7, 1906, JD File 46330-84974; Mrs. Bliss Black to M. D. Purdy, July 19, 1906, JD File 46330-85881. Frank B. Kellogg and Charles Morrison, "Report and Opinion upon the Legality of the Standard Oil Company Under the Sherman Act," p. 1, JD File 46330, box 160. Roosevelt to Moody, September 13, 1906, in Morison, *Letters of Roosevelt,* 5:409.

83. *United States* v. *Standard Oil Company of New Jersey,* 173 Fed. 177 (1909), *Record* (Washington: Government Printing Office, 1909), 23:111.

84. On Milburn, see *National Cyclopaedia of American Biography* (New

York: James T. White and Company, 1898-1951), 32:426; on Johnson, *Dictionary of American Biography* (New York: Charles Scribner's Sons, 1928-1936), 10:106; on Watson, *DAB*, 19:540.

85. Kellogg discussed this move by the defense attorneys in a letter to Bonaparte, January 8, 1907, JD File 46330-97374.

86. *United States* v. *Standard Oil Co. of New Jersey, et al.*, 152 Fed. 290, 292, 297 (1907). On Bonaparte, see Allen W. Rumble, "Rectitude and Reform: Charles J. Bonaparte and the Politics of Gentility, 1851-1921" (Ph.D. diss., University of Maryland, 1971); Joseph B. Bishop, *Charles J. Bonaparte* (New York: Charles Scribner's Sons, 1922); and Eric F. Goldman, *Charles J. Bonaparte, Patrician Reformer: His Earlier Career* (Baltimore: Johns Hopkins University Press, 1943).

87. *U.S.* v. *Standard Oil, Record,* 23 vols. On the role of the Bureau of Corporations, see Morrison to H. K. Smith, April 16, 1909, Bureau of Corporations, File 4167-5-11, Record Group 122, National Archives, Washington, D.C. Also see Morrison to Dana Durand, April 8, 1908, BC File 4167-5-4.

88. Bonaparte to Roosevelt, August 22, 1907, box 183, Bonaparte Papers, Manuscript Division, Library of Congress; Kellogg to Purdy, June 18, 1907, JD File 46330-110519.

89. Purdy to Moody, July 18, 1906, BC File 4167-5, pt. 1.

90. Bonaparte to Roosevelt, August 22, 1907, box 183, Bonaparte Papers.

91. Ibid., September 8, 1907, box 183, and Roosevelt to Bonaparte, September 10, 1907, box 157, Bonaparte Papers.

92. Standard Oil Company to Bonaparte, September 29, 1907, box 131, Bonaparte Papers; Garfield, 1907 diary, September 21, 22, 28, 29, box 8, Garfield Papers.

93. Kellogg to Bonaparte, October 6, 1907, JD File 46330-118158.

94. Morrison to Bonaparte, October 22, 1907, JD File 46330-119460; Garfield, 1907 Diary, October 25, Box 8, Garfield Papers, LC; Bonaparte to Standard Oil Company, October 25, 1907, box 131, Bonaparte Papers.

95. Roosevelt to Bonaparte, July 25, 1908, Bonaparte Papers, box 132, Bonaparte Papers.

96. Perkins's message is quoted in Kellogg to Bonaparte, November 23, 1907, JD File 46330-122149.

97. Bonaparte to Kellogg, November 26, 1907, JD File 46330-122149.

98. Kellogg to Bonaparte, December 2, 1907, and Bonaparte to Kellogg, December 3, 1907, JD File 46330-122685.

99. *United States* v. *Standard Oil Company of Indiana,* 155 Fed. 305 (1907); Paul H. Giddens, *The Standard Oil Company (Indiana)* (New York: Appleton-Century-Crofts, 1955), pp. 100-21, is the fullest account of the Landis decision.

100. Archbold to Bonaparte, April 8, 1908, and Bonaparte to Archbold, April 28, 1908, box 6, Bonaparte Papers, Maryland Historical Society, Baltimore.

101. Edwin W. Sims to Bonaparte, May 6, 1908, box 6, Bonaparte Papers, MHS.

102. Bonaparte to Archbold, n.d. (May 19, 1908), and Archbold to Bonaparte, May 28, 1908, box 6, Bonaparte Papers, MHS.

103. Archbold to Bonaparte, June 4, 1908, box 6, Bonaparte Papers, MHS.

104. Roosevelt to Bonaparte, June 26, 29, July 25, 1908, box 132, and Bonaparte to Roosevelt, July 3, 1908, box 184, Bonaparte Papers, LC. Archbold to Bonaparte, July 20, 1908, box 6, Bonaparte Papers, MHS. The party was covered in the *Boston Journal,* July 27, 1908.

105. Roosevelt, "Special Message," April 14, 1908, in James D. Richardson, ed., *The Messages and Papers of Presidents* (New York: Bureau of National Literature, 1903-1912), 14:7138, 7146.

106. Morison, *Letters of Roosevelt,* 6:1243, 1249.

107. Rockefeller told reporters: "I expect to cast my vote for William Howard Taft. If for no other reason, I support Mr. Taft because on comparing him with Mr. Bryan, his chief opponent, I find a balance of fitness and temperament entirely on his side. The election of Mr. Taft will, I believe, make for law and order and stability of business." *New York Times,* October 30, 1908. Roosevelt and Taft's statements both appeared in ibid., October 31, 1908.

CHAPTER 6

1. "Four Aspects of Civic Duty," quoted in Alpheus T. Mason, *William Howard Taft: Chief Justice* (London: Oldbourne Book Company, 1964), p. 50; "Special Message," January 7, 1910, in James D. Richardson, comp., *The Messages and Papers of Presidents* (New York: Bureau of National Literature, 1903-1912), 15:7449.

2. "Address to the Pocatello Chamber of Commerce," quoted in Mason, *Taft,* p. 19.

3. Archibald W. Butt, *Taft and Roosevelt: The Intimate Letters of Archie Butt* (Garden City, N.J.: Doubleday, Doran, 1930), 1:110-11.

4. Taft to Otto Bannard, June 29, 1912, series 8, vol. 38, Taft Papers, Manuscript Division, Library of Congress.

5. "Annual Message," December 6, 1910, in Richardson, *Messages and Papers,* 15:7554-55.

6. United States, Library of Congress, Legislative Reference Service, *Congress and the Monopoly Problem* (Washington: Government Printing Office, 1956), p. 659.

7. George Wickersham, "Recent Interpretation of the Sherman Act," *Michigan Law Review* 10 (November 1911): 24. Also see Charles Johnson, "The Attorney-General and the Trusts," *Harper's Weekly* 55 (May 22, 1911): 8.

8. *New York Times,* May 1, 1909; *The Outlook,* 41 (March 27, 1909): 627; *Independent* 66 (March 11, 1909): 513; James C. German, "Taft's Attorney General: George W. Wickersham" (Ph.D. diss., New York University, 1969), pp. 1-6.

9. *Congressional Record* 44, pt. 2, 61st Cong., 1st sess., p. 1935.

10. *Wall Street Journal,* September 28, 1911.

11. *Literary Digest* 40 (January 1, 1910): 4.

12. Wickersham to Taft, February 7, 1911, series 6, file 778, Taft to Wickersham, February 9, 1911, series 8, vol. 23, Taft Papers; Henry F. Pringle, *The Life and Times of William Howard Taft* (New York: Toronto, Farrer & Rinehart, 1939), 2:668.

13. Butt, *Taft and Roosevelt,* 2:615-16.

14. *Literary Digest* 38 (April 17, 1909): 631-33. For brief sketches of the members of the Court, see *Current Literature* 48 (January 1910): 15-20; *Literary Digest* 39 (December 4, 1909): 991-93.

15. *United States* v. *Northern Securities Company,* 120 Fed. 721 (1903), on appeal, 193 U.S. 197 (1904).

16. *U.S.* v. *Standard Oil of New Jersey,* 173 Fed. 177 (1909), *Brief on Law for Petitioner,* p. 160.

17. *Brief on Law for Petitioner,* p. 194.

18. *Northern Securities Company* v. *United States,* 193 U.S. 197, 410 (1904).

19. *U.S.* v. *Standard Oil, Defendant's Brief,* 4:71-79.

20. Ibid., 5:151-57.

21. *U.S.* v. *Standard Oil,* at 189-90.

22. Ibid., at 191.

23. On the justices' comparative lack of economic expertise, see the following sketches: Sanborn in *Dictionary of American Biography* (New York: Charles Scribner's Sons, 1928-1936), 16:328; Adams in *National Cyclopaedia of American Biography* (New York: James T. White and

Company, 1898-1951), 5:385-86; Hook in *Who's Who in America, 1908-1909,* ed. Albert N. Marquis (Chicago: A. N. Marquis & Co., 1909), p. 915; and David Burner, "Willis Van Devanter," in Leon Friedman and Fred L. Israel, eds., *Justices of the United States Supreme Court* (New York: R. R. Bowker Company, 1969), 3:1945-54.

24. *U.S.* v. *Standard Oil,* at 189.

25. Ibid., at 199.

26. See list of leading stockholders in *The Independent* 63 (September 26, 1907): 715.

27. *U.S.* v. *Standard Oil,* at 182.

28. Ibid., at 199-200.

29. Ibid., at 192.

30. Kellogg to Wickersham, November 20, 1909, JD File 46330-89, Record Group 60, National Archives, Washington, D.C.

31. *New York Times,* November 21, 1909.

32. Taft to Kellogg, November 21, 1909, series 8, vol. 8, Taft Papers; Butt, *Taft and Roosevelt,* 2:221.

33. On Harlan, see Alan F. Westin, "John Marshall Harlan," in Allison Dunham and Philip B. Kurland, eds., *Mr. Justice* (Chicago: University of Chicago Press, 1962), pp. 98-116; Louis Filler, "John M. Harlan," in Friedman and Israel, *Justices of the Supreme Court,* 2:1291-95; *DAB,* 8:269-73. For Harlan's defense of the Sherman Antitrust Act, see Floyd B. Clark, *The Constitutional Doctrines of Justice Harlan* (Baltimore: Johns Hopkins University Press, 1915); Loren P. Beth, "Justice Harlan and the Uses of Dissent," *American Political Science Review* 49 (December 1955): 1080-1104. On McKenna, see Matthew McDevitt, *Joseph McKenna* (Washington, 1946); James F. Watts, "Joseph McKenna," in Friedman and Israel, *Justices of the Supreme Court,* 3:1719-36; *DAB,* 12:87-89. On Day, see Joseph E. McLean, *William Rufus Day* (Baltimore: Johns Hopkins University Press, 1946); James F. Watts, "William R. Day," in Friedman and Israel, *Justices of the Supreme Court,* 3:1773-89; *DAB,* 5:163-65. On Van Devanter, see David Burner, "Willis Van Devanter," in Friedman and Israel, *Justices of the Supreme Court,* 3:1945-53; German, "Wickersham," p. 58, discusses the attorney general's role in Van Devanter's appointment to the Supreme Court.

34. Holmes to Pollock, April 23, 1910, in Mark DeWolfe Howe, ed., *Holmes-Pollock Letters* (Cambridge: Harvard University Press, 1941), 1:163. Brief accounts of Holmes's career are Francis Biddle, "Mr. Justice Holmes," in Allison Dunham and Phillip Kurland, eds., *Mr. Justice* (Chicago: University of Chicago Press, 1962), pp. 1-14; Paul A. Freund, "Oliver Wendell Holmes," in Friedman and Israel, *Justices of the Supreme*

Court, 3:1755-62; and Felix Frankfurter, "Oliver Wendell Holmes," *DAB,* 21:417-27. Mark DeWolfe Howe, *Justice Oliver Wendell Holmes* (Cambridge: Harvard University Press, 1957-1962) is the definitive account of the Justice's early years. On Chief Justice White, see Marie C. Klinkhamer, *Edward Douglas White* (Washington, 1943); Robert B. Dishman, "Mr. Justice White and the Rule of Reason," *The Review of Politics* 41 (1951): 229-32; James F. Watts, "Edward Douglas White," in Friedman and Israel, *Justices of the Supreme Court,* 3:1633-59.

35. On Hughes' early career, see Merlo Pusey, *Charles Evans Hughes* (New York: Macmillan Company, 1951); Robert F. Wesser, *Charles Evans Hughes: Politics and Reform in New York, 1905-1910* (Ithaca: Cornell University Press, 1967); and David J. Danelski and Joseph S. Tulchin, eds., *The Autobiographical Notes of Charles Evans Hughes* (Cambridge: Harvard University Press, 1973). On Lamar, see Clarinda P. Lamar, *The Life of Joseph Rucker Lamar* (New York: G. P. Putnam's Sons, 1926); Leonard Dinnerstein, "Joseph Rucker Lamar," in Friedman and Israel, *Justices of the Supreme Court,* 3:1973-89; *DAB* 10:550-51. On Lurton, see James F. Watts, "Horace H. Lurton," in Friedman and Israel, *Justices of the Supreme Court,* 3:1847-63; *DAB,* 11:509-10.

36. *Standard Oil Company of New Jersey* v. *United States,* 221 U.S. 1 (1911), *Brief for the United States,* no. 725. These two volumes closely parallel the briefs submitted to the circuit court. Frank B. Kellogg, *Standard Oil Company of New Jersey* v. *United States . . . Oral Argument of Frank B. Kellogg on behalf of the United States* (Washington: Government Printing Office, 1911).

37. *Brief for the Standard Oil Company,* no. 725; John G. Milburn, *Standard Oil Company of New Jersey* v. *United States . . . Oral Arguments on Behalf of Appellants* (New York: C. G. Gurgoyne, 1911). See summary of defense presentation in *The Outlook* 44 (March 26, 1910): 645.

38. Edward G. Lowry, "The Supreme Court Speaks," *Harper's Weekly* 55 (June 3, 1911): 8; also see the account of the decision in *New York Times, New York World, New York Tribune,* and New York *Sun,* all May 16, 1911. Kellogg to Wickersham, March 17, April 3, 11, 1911, JD File 60-57-0, sec. 2.

39. Lowry, "Supreme Court Speaks," p. 8.

40. *Standard Oil Company of New Jersey* v. *United States,* 221 U.S. 1 (1911), p. 48.

41. Ibid., at 75.

42. Ibid., at 80-81.

43. *John Dyer's Case,* Y.B. 2 Hen. V, f. 5, pl. 26 (1415), is the most important decision concerning early contracts in restraint of trade. For an examination of this and other cases, see Herman Oliphant, *Cases on Trade*

Regulation Selected from Decisions of English and American Courts (St. Paul: West Publishing Company, 1923), pp. 34ff; Hans B. Thorelli, *Federal Antitrust Policy* (Baltimore: Johns Hopkins University Press, 1955), p. 17; "Notes," *Columbia Law Review* 32 (February 1932): p. 292. William Letwin, *Law and Economic Policy in America* (New York: Random House, 1965), pp. 39-40 minimizes the importance of Dyer's case. For the gradual relaxation of public policy against restraints of trade in the seventeenth century, see *Mitchell* v. *Reynolds,* 1 P. Wms. 181, 24 Eng. Rep. 347 (1711); *Rogers* v. *Parrey,* 2 Bulst. 136, 80 Eng. Rep. 1012 (1613); and *Broad* v. *Jollyfe,* Cro. Jac. 596, 79 Eng. Rep. 509 (1620).

44. *Nordenfelt* v. *Maxim Nordenfelt Gun Co.,* App. Cas. 535 (1834).

45. *Collins* v. *Locke,* 4 L.R.A.C. 674, 678 (1879).

46. *Central Ohio Salt Company* v. *Guthrie,* 35 Ohio 666, 672 (1880). Also see *Craft et al.* v. *McConoughy,* 79 Ill. 346 (1875); *India Bagging Assn.* v. *B. Kock & Co.,* 14 La. 168 (1859); *Emery* v. *Ohio Candle Co.,* 47 Ohio 320 (1890).

47. Milton Handler, *A Study of the Construction and Enforcement of Federal Antitrust Law,* Temporary National Economic Committee monograph no. 38 (Washington: Government Printing Office, 1941), p. 4.

48. *United States* v. *Trans-Missouri Freight Association,* 166 U.S. 290 (1897). Benjamin Twiss, *Lawyers and the Constitution* (Princeton: Princeton University Press, 1942), pp. 190-92, traces the contributions of railroad attorneys to the development of the legal doctrine of reasonable restraints.

49. *Trans-Missouri,* at 328.

50. See *Lochner* v. *New York,* 198 U.S. 45 (1905).

51. *Trans-Missouri,* at 324.

52. Ibid., at 346.

53. Ibid., at 355.

54. White's papers have not survived. The secondary sources, cited in note 35, fail to account adequately for White's devotion to the rule of reason.

55. White rejected the substantive due process doctrine and voted to uphold state legislation regulating the length of the working day in *Lochner* v. *New York,* 198 U.S. 45 (1905), and *Muller* v. *Oregon,* 208 U.S. 412 (1908). But he changed position on the same issue in *Bunting* v. *Oregon,* 243 U.S. 426 (1917). Similarly, White supported an expanded view of the federal commerce power in *Champion* v. *Ames,* 188 U.S. 321 (1903); *McCray* v. *U.S.,* 195 U.S. 27 (1904); *Hipolite Egg Co.* v. *U.S.,* 220 U.S. 45 (1911); and *Hoke* v. *U.S.,* 227 U.S. 308 (1913). But he switched position on the scope of the commerce power in *Hammer* v. *Dagenhart,* 247 U.S. 251 (1918).

56. *United States* v. *Joint Traffic Association,* 171 U.S. 505 (1898). The railroad attorney's argument is quoted in Benjamin Twiss, *Lawyers and the Constitution* (Princeton: Princeton University Press, 1942), p. 194.

57. *Joint Traffic Association,* at 568.

58. *Hopkins et al.* v. *United States,* 171 U.S. 572, 602 (1898).

59. *Anderson et al.* v. *United States,* 171 U.S. 604, 616 (1898).

60. *United States* v. *Addystone Pipe & Steel Company,* 85 Fed. 271 (1898).

61. Ibid., at 282-83.

62. *Addystone Pipe & Steel Company* v. *United States,* 175 U.S. 211 (1899).

63. *Northern Securities,* at 405.

64. Ibid., at 361.

65. S. 3937 (introduced January 29, 1904), *Congressional Record* 38, pt. 2, 58th Cong., 2d sess., pp. 1360-61; S. 9331 (introduced March 25, 1908). *Congressional Record,* 43, pt. 4, 60th Cong., 1st sess., p. 3852.

66. Arthur M. Johnson, "Antitrust Policy in Transition, 1908," *Mississippi Valley Historical Review* 48 (1961): 415-34, traces the evolution of the Hepburn Bill of 1908.

67. United States, House of Representatives, Committee on Judiciary, *Hearings on House Bill 19745,* 60th Cong., 1st sess. (Washington: Government Printing Office, 1908), Smith testimony, pp. 397-411; Jencks testimony, pp. 108-22; Low testimony, p. 16.

68. Quote from *Senate Report no. 848,* 60th Cong., 2d sess., p. 11. For Senate action on Warner's bill (S.6440), see *Congressional Record* 42, pt. 5, 60th Cong., 1st sess., pp. 4212, 4395; 43, pt. 2, 60th Cong., 2d sess., p. 1395.

69. Roosevelt to S. Low, April 1, 1908, in Elting E. Morison, ed., *The Letters of Theodore Roosevelt* (Cambridge: Harvard University Press, 1951-1954), 6:986-87.

70. Richardson, *Messages and Papers* 15: 7454.

71. *Standard Oil Company of New Jersey* v. *United States,* 221 U.S. 1, 56 (1911).

72. Ibid., at 63.

73. *Brief for the United States,* 1:297ff.

74. *SONJ* v. *U.S.* at 64-65. Italics added.

75. Ibid., at 66.

76. Reported in *New York Times,* May 16, 1911. The animosity between Harlan and White had a long history. Justice Charles Evans Hughes recalled White's account of the first meeting between White and Harlan: "Harlan had been sent to Louisiana by President Hayes, in the spring of

1877, as a member of the Commission to bring about a settlement of local controversies. The Commission was met by a delegation of which White was either the head or a member. White said that he described freely to the Commission the actual conditions in the State, and when he was through a very large man (Harlan) with a buff-colored vest rose from his seat and approached White and said, 'Well, you are damned frank.''' Danelski and Tulchin, *Autobiographical Notes of Hughes,* p. 170.

77. *SONJ* v. *U.S.* at 83.

78. Ibid., at 96.

79. Ibid., at 97.

80. Ibid., at 103.

81. Ibid., at 83.

82. Holmes to Pollock, September 10, 1910, in Howe, ed., *Holmes-Pollock Letters,* 1:170.

83. Charles Evans Hughes, *The Supreme Court of the United States* (New York: Garden City Publishing Company, 1928), p. 231.

84. Pusey, *Hughes,* 1:282.

85. Danelski and Tulchin, eds., *Autobiographical Notes of Hughes,* p. 168.

86. Ibid., p. 170.

87. *United States* v. *American Tobacco,* 221 U.S. 106, 108 (1911).

88. Ibid., at 187.

89. Ibid., at 190.

90. On the effect of the tobacco dissolution, see Richard B. Tennant, *The American Cigarette Industry* (New Haven: Yale University Press, 1950), pp. 64-65; Henry R. Seager and Charles A. Gulick, *Trust and Corporation Problems* (New York: Harper and Brothers, 1929), pp. 173-78; Eliot Jones, *The Trust Problem in the United States* (New York: Macmillan, 1926), pp. 453-74.

91. Taft's New Haven address on June 21, 1911, quoted by Senator La Follette in *Congressional Record* 47, pt. 5, 62d Cong., 1st sess., p. 4188; Richardson, *Messages and Papers,* 15:7644.

92. Taft to Helen H. Taft, May 16, 1911, series 2, Taft Papers, LC.

93. *New York Times,* May 16, 1911; Wickersham, "Recent Interpretation of the Sherman Act," p. 23; Walker to Wickersham, June 6, 1911, JD File 60-57-0, sec. 2; Walker to Wickersham, June 21, 1911, and Wickersham to Walker, June 26, 1911, JD File 60-57-0, sec. 3.

94. *The Outlook* 98 (June 3, 1911): 239-40.

95. *New York Times,* May 16, 1911.

96. *Congressional Record* 47, pt. 5, 62d Cong., 1st sess., pp. 4183-93.

97. *Current Literature* 51 (July 1911): 1.

98. S. 2374, S. 2375, S. 2433, H.R. 10508, H.R. 11855, H.R. 13003, *Congressional Record* 47, pts. 2-4, 62d Cong., 1st sess., pp. 1267, 1299, 1530, 2305, 3287; Kirk H. Porter and Donald B. Johnson, *National Party Platforms,* 2d ed. (Urbana: University of Illinois Press, 1961), p. 169.

99. Wilson's acceptance speech contained only vague moralizing on the trust issue. Ray Stannard Baker and William E. Dodd, eds., *The Public Papers of Woodrow Wilson* (New York: Harper and Brothers, 1926), 2:455-56. On the development of Wilson's position on trusts, see Alpheus T. Mason, *Louis D. Brandeis: A Free Man's Life* (New York: Viking Press, 1946), p. 353; Belle C. and Fola La Follette, *Robert M. La Follette* (New York: Macmillan, 1953), 1:336-37; Arthur S. Link, *Wilson: The Road to the White House* (Princeton: Princeton University Press, 1947), pp. 490-93.

100. United States Senate, Committee on Interstate Commerce, *Report . . . Pursuant to Senate Resolution 98,* 62d Cong., 3d sess., S. Rept. 1326 (Washington: Government Printing Office, 1913), p. xii.

101. Richardson, *Messages and Papers,* 17:7913-19.

102. *Congressional Record* 51, pt. 9, 63d Cong., 2d sess., p. 9270.

103. Arthur S. Link, *Wilson: The New Freedom* (Princeton: Princeton University Press, 1956), pp. 425-27.

104. Ibid., p. 443.

105. Quoted in ibid., p. 444.

106. *United States* v. *United Shoe Machinery Company,* 247 U.S. 32, 56 (1918). Justices McReynolds and Brandeis did not participate in this decision.

107. Ibid., at 75.

108. Ibid., at 90.

109. *United States* v. *United States Steel Corporation,* 251 U.S. 417, 451 (1920). Justices McReynolds and Brandeis did not participate in this decision.

110. Ibid., at 461.

111. The government dismissed appeals in the following cases: *United States* v. *Quaker Oats Co.,* 232 Fed. 499 (1916), 253 U.S. 499 (1920); *United States* v. *American Can Co.,* 230 Fed. 859 (1916), 234 Fed. 1019 (1916), 256 U.S. 706 (1921); *United States* v. *Keystone Watch Co.,* 218 Fed. 502 (1915), 257 U.S. 664 (1921).

112. *United States* v. *International Harvester Co.,* 274 U.S. 693, 703, 708 (1927). Justice Brandeis did not participate in this decision.

113. Harry L. Purdy, Martin Lindahl, and William Carter, *Corporate Concentration and Public Policy,* 2d ed. (New York: Prentice-Hall, 1950), pp. 27-29; M. A. Adelman, "Effective Competition and the Anti-Trust Laws," *Harvard Law Review* 61 (September 1948): 1293, n. 8.

CHAPTER 7

1. M.F. Elliot to J.O. Slonecker, July 12, 1911, quoted in George S. Gibb and Evelyn H. Knowlton, *The Resurgent Years* (New York: Harper and Brothers, 1956), p. 15.

2. Albert W. Atwood, "The Greatest Killing on Wall Street," *McClure's* 39 (August 1912): 417; Gibb and Knowlton, *The Resurgent Years,* pp. 22-23; Ralph Hidy and Muriel Hidy, *Pioneering in Big Business* (New York: Harper and Brothers, 1955), p. 699; Peter Collier and David Horowitz, *The Rockefellers* (New York: Signet, 1977), p. 65.

3. Henry H. Rogers, interview, *New York Times,* January 7, 9, 10, 11, 13, 14, 1906; John D. Archbold, "To the Press and Public," *New York Times,* September 26, 27, 1907; Archbold, "The Standard Oil Company," *Saturday Evening Post* 180 (December 7, 1907): 3-5; Leonard Woolsey Bacon, *History of the South Improvement Company* (New York, 1907), noted in Hidy and Hidy, *Pioneering in Big Business,* p. 702; *World's Work,* 16 (October 1908): 10755-68; John D. Rockefeller, *Random Reminiscences of Men and Events* (New York: Doubleday, Page & Co., 1909).

4. Harold F. Williamson et al., *The American Petroleum Industry: The Age of Energy* (Evanston: Northwestern University Press, 1963), p. 16.

5. Ralph Andreano, "The Emergence of New Competition in the American Petroleum Industry Before 1911" (Ph.D. diss., Northwestern University, 1960), appendix A, pp. 306-10, discusses flush fields.

6. Ibid., pp. 68-72, 105-07, 158-65, 221-27, 242-44, for the impact of varying types of crude on the competitive structure of the industry.

7. Ibid., pp. 67, 110-19, 228-32, 244-51.

8. Frank B. Kellogg, "Results of the Standard Oil Decision," *American Review of Reviews* 45 (June 1912): 728; Wickersham's statement in *Literary Digest* 44 (June 15, 1912): 1241.

9. Milburn to Wickersham, June 9, 1911, JD File 60-57-0, sec. 2, Record Group 60, National Archives, Washington, D.C.

10. Ibid., June 13, 1911, JD File 60-57-0, sec. 3.

11. Kellogg to Wickersham, June 14, 1911, JD File 60-57-0, sec. 3.

12. Wickersham to Kellogg, June 21, 1911, JD File 60-57-0, sec. 3.

13. Hidy and Hidy, *Pioneering in Big Business,* p. 711.

14. Hadley to Wickersham, November 29, 1909, JD File 46330-102.

15. Wickersham to Hadley, December 2, 1909, JD File 46330-102.

16. JD File 60-57-0, sec. 2, contains many letters urging criminal prosecutions of the Standard Oil leaders.

17. Senate Resolution, May 23, 1911, JD File 60-57-0, sec. 2; Wickersham to the President of the Senate, May 25, 1911, JD File 60-57-0, sec. 2.

18. *The Outlook* 98 (August 12, 1911): 804.

19. *Literary Digest* 43 (August 12, 1911): 226-28.

20. New York *World,* September 7, 1911.

21. *Literary Digest,* 44 (March 23, 1912): 576.

22. Paul H. Giddens, *Standard Oil Company (Indiana)* (New York: Appleton-Century-Crofts, 1955), p. 137.

23. Carl H. Pforzheimer, *Standard Oil Issues* (November 1912), quoted in Allan Nevins, *John D. Rockefeller: The Heroic Age of American Enterprise,* vol. 2 (New York: Charles Scribner's Sons, 1940), p. 607.

24. Atwood, "Greatest Killing on Wall Street," pp. 409-413.

25. Ibid., p. 419; Giddens, *Standard Oil Company (Indiana),* p. 131.

26. *Literary Digest* 45 (November 30, 1912): 1037-38.

27. "What's the Explanation?" *Everybody's Magazine,* 26 (June 1912): 817-18.

28. See press comments collected in *Literary Digest* 44 (June 15, 1912): 1240.

29. Quoted in Ibid.

30. *New York Times,* July 22, 1913; Tom Finty, Jr., *Anti-Trust Legislation in Texas* (Galveston, Tex.: A.H. Belo & Company, 1916), p. 66.

31. J.H. Swift to Wickersham, October 16, 1911, JD File 60-57-0, sec. 3; P.F. Mansfield to Wickersham, December 31, 1911, JD File 60-57-0, sec. 4; P.G. Peabody to Wickersham, August 23, 1911, JD File 60-57-0, sec. 3; H.J. Melcher to H.K. Smith, February 20, 1912, JD File 60-57-0, sec. 4; C.P. Anderbery to Justice Department, July 1, 1912, JD File 60-57-0, sec. 4; P.G. Peabody to Wickersham, August 3, 1911, JD File 60-57-0, sec. 3.

32. W.R. Adams to Wickersham, March 19, 1912, JD File 60-57-0, sec. 4.

33. Owen to Taft, December 6, 1911, JD File 60-57-0, sec. 4; Wickersham to Owen, December 15, 1911, JD File 60-57-0, sec. 4; Owen to Wickersham, January 2, 1912, JD File 60-57-0, sec. 4.

34. Washington *Times,* September 29, 1912.

35. Wickersham to Kellogg, October 21, 1912, JD File 60-57-0, sec. 5.

36. Wickersham to Milburn, November 10, 1911, JD File 60-57-0, sec. 3; Milburn to Wickersham, November 13, 17, 20, 1911, JD File 60-57-0, sec. 4.

37. Wickersham to George Jonas, May 6, 1912, JD File 60-57-0, sec. 4.

38. Wickersham to Morrison, October 17, 1912, JD File 60-57-0, sec. 5.

39. Morrison to Assistant Attorney General Fowler, November 15, 1912, JD File 60-57-0, sec. 5.

40. "Report of C.B. Morrison, Special Assistant to the Attorney

General, upon the Oil Investigation," February 28, 1913, pp. 14, 79, 97, JD File 60-57-0, sec. 7.

41. JD File 60-57-0, sec. 6, contains many complaints against Standard Oil received during the first months of the Wilson administration.

42. On McReynolds, see *Dictionary of American Biography* (New York: Charles Scribner's Sons, 1928–1936), sup. IV, pp. 536-38; David Burner, "James C. McReynolds," in Leon Friedman and Fred L. Israel, eds., *Justices of the United States Supreme Court* (New York: R.R. Bowker Company, 1969), 3: 2023-33.

43. Morrison to Fowler, May 19, 1913, JD File 60-57-0, sec. 6.

44. Morrison to McReynolds, June 17, 1913, JD File 60-57-0, sec. 6.

45. Assistant Attorney General Todd to Morrison, January 19, 1913, JD File 60-57-0, sec. 6; Morrison to McReynolds, February 16, 1914, JD File 60-57-0, sec. 7.

46. McReynolds to Milburn, July 9, 1913, JD File 60-57-0, sec. 6.

47. Fowler to Homer M. Boardman, July 19, 1913, JD File 60-57-0, sec. 6.

48. Chicago *Examiner,* August 26, 1914.

49. Morrison to Todd, August 26, 1914, JD File 60-57-0, sec. 8; JD File 60-57-0, sec. 8, contains a copy of the Senate resolution of August 27, 1914.

50. McReynolds to J.P. Tumulty, August 28, 1914, JD File 60-57-0, sec. 8.

51. On Gregory, see *DAB,* sup. I, pp. 358-60.

52. Morrison, "Progress Report on Standard Oil Investigation," October 30, 1914, p. 13, JD File 60-57-0, sec. 9.

53. Morrison to Todd, June 28, 1915, JD File 60-57-0, sec. 9.

54. Todd to Morrison, July 7, 1915, JD File 60-57-0, sec. 9; Morrison to Todd, November 27, 1915, JD File 60-57-0, sec. 9; Todd to Morrison, December 13, 1915, JD File 60-57-0, sec. 9; Todd to A.A. Herring, January 4, 1916, JD File 60-57-0, sec. 9.

55. *Congressional Record* 53, pt. 6, 64th Cong., 1st sess., p. 5870.

56. Gregory to the President of the Senate, April 12, 1916, JD File 60-57-0, sec. 10.

57. Morrison to Attorney General, June 6, 1916, JD File 60-57-0, sec. 10.

58. Morrison to Todd, June 17, 1916, JD File 60-57-0, sec. 10.

59. Ibid., July 5, 1917, JD File 60-57-0, sec. 10.

60. Ibid.

61. "Memorandum to Be Attached to Mr. Morrison's Letter of July 5, 1917, in re Standard Oil Investigation," JD File 60-57-0, sec. 10.

62. As Arthur S. Link observed, "There was no trust-busting ardor in

the Wilson administration." *Woodrow Wilson and the Progressive Era* (New York: Harper & Row, 1954), p. 76.

63. *Reader's Guide to Periodical Literature,* vols. 2-6, supps. 1, 2.

64. "Pioneers of Standard Oil," *Nation* 103 (December 14, 1916): 556; "Fifty Years of the Standard Oil Company," *World's Work* 39 (March 20, 1920): 431-32; "At Thirty-Nine He Became a Great Captain of Industry," *Current Opinion* 64 (January 1918): 21-22. "Good News for Oil and Steel Workers," *The Outlook* 118 (April 17, 1918): 612; "Standard Oil and the Eight Hour Day," *The Outlook* 111 (September 15, 1915): 109-10; "Those Who Can Afford Welfare Work," *World's Work* 36 (June 1918): 131-32; "Standard Oil and China, Partners," *Literary Digest* 48 (February 28, 1914): 421-22; "New Light of Asia a General Convenience," *Literary Digest* 55 (November 17, 1917): 79; E.S. Yale, "Carriers of Light and Laughter," *World Outlook* 3 (November 1917): 14-15.

65. Quoted in Williamson et al., *Age of Energy,* p. 261.

66. Ibid., pt. 2, contains a good overview of the evolution of the petroleum industry into the 1920s.

67. Federal Trade Commission, *Report on the Pipe Line Transportation of Petroleum* (Washington: Government Printing Office, 1916), p. 4.

68. Federal Trade Commission, *Report on the Price of Gasoline in 1915* (Washington: Government Printing Office, 1917), p. 164.

69. United States Senate, Committee on Manufactures, *High Cost of Gasoline and Other Petroleum Products,* 67th Cong., 4th sess., Senate Report 1269 (Washington: Government Printing Office, 1923), p. 48.

70. Federal Trade Commission, *Petroleum Industry: Prices, Profits and Competition,* 70th Cong., 1st sess., Senate Doc. no. 61 (Washington: Government Printing Office, 1928), p. 42.

71. Federal Trade Commission, *Report on the Pacific Coast Petroleum Industry* (Washington: Government Printing Office, 1921), pt. 2, pp. 30–31.

72. Federal Trade Commission, *Petroleum Industry,* pp. 193-202.

73. Federal Trade Commission, *Report on the Price of Gasoline in 1915,* pp. 7, 113.

74. United States Fuel Administration, *Prices and Marketing Practices Covering the Distribution of Gasoline and Kerosene Throughout the United States,* by A.G. Maguire (Washington: Government Printing Office, 1919), p. 11.

75. United States Senate, *High Cost of Gasoline,* pp. 162, 778-79.

76. Fuel Administration, *Prices and Marketing Practices,* pp. 7-11.

77. Federal Trade Commission, *Report on the Price of Gasoline in 1915,* pp. 143-45; Gibb and Knowlton, *Resurgent Years,* pp. 778–79.

78. Federal Trade Commission, *Report on the Price of Gasoline in 1915,* pp. 5-6.

79. Federal Trade Commission, *The Advance in the Price of Petroleum Products* (Washington: Government Printing Office, 1920), p. 53.

80. Federal Trade Commission, *Petroleum Industry,* pp. 228-29.

81. Gibb and Knowlton, *Resurgent Years,* pp. 110-35, 547; Williamson et al., *Age of Energy,* pp. 136–62, 374-94.

82. Federal Trade Commission, *Report on the Price of Gasoline in 1915,* p. 5.

83. United States Senate, *High Cost of Gasoline,* p. 3.

CHAPTER 8

1. Henry Demarest Lloyd, *Wealth Against Commonwealth* (Washington: National Home Library Association, 1936), p. 536.

2. Originally an advocate of antitrust laws, Lloyd lost faith in such legislation early in the 1890s. Chester M. Destler, *Henry Demarest Lloyd and the Empire of Reform* (Philadelphia: University of Pennsylvania Press, 1963), pp. 181, 238.

3. William Letwin, *Law and Economic Policy in America* (New York: Random House, 1965), pp. 71-85. See S.C.T. Todd, "The Present Legal Status of Trusts," *Harvard Law Review* 7 (October 1893): 157-65, for a Standard Oil lawyer's view on the inevitability of trusts.

4. Ralph Andreano, "The Emergence of New Competition in the American Petroleum Industry Before 1911" (Ph.D. diss., Northwestern University, 1960); Harold F. Williamson et al., *The American Petroleum Industry: The Age of Energy* (Evanston: Northwestern University Press, 1963), pts. 1, 2.

5. Guenter Reimann, "Standard Oil and I.G. Farben," *New Republic* 105 (August 4, 1941): 147-49; Eliot Janeway, "Economic Warfare," *Fortune* 24 (August 1942): 101, 118-22. See Gabriel Kolko, "American Business and Germany, 1930-41," *Western Political Quarterly* 15 (December 1962): 713-28, for a survey of American business ties with Nazi Germany.

6. United States Attorney General, *Annual Report, 1941* (Washington: Government Printing Office, 1942), pp. 58-63; Ellis Hawley, *The New Deal and the Problem of Monopoly* (Princeton: Princeton University Press, 1964), pp. 441-43.

7. United States Senate, Committee Investigating the National Defense Program, *Hearings Before a Special Committee Investigating the National*

Defense Program, 77th Cong., 1st sess. (Washington: Government Printing Office, 1942), pt. 11, exhibits 360-66, pp. 4561-84.

8. United States Senate, Committee on Patents, *Hearings Before the Committee on Patents,* 77th Cong., 2nd sess. (Washington: Government Printing Office, 1942), exhibit 44, p. 2904.

9. Ibid., exhibit 47, p. 2906.

10. *Investigation of the National Defense Program,* pt. 11, exhibit 381, p. 4602.

11. *Hearings Before the Committee on Patents,* exhibit 165, pp. 2722-46, exhibit 230, p. 3143.

12. *Investigation of the National Defense Program,* exhibit 366, p. 4583.

13. *United States* v. *Standard Oil Co., et al.,* in U.S. District Court for the District of New Jersey, November Term, 1941, reprinted in *Investigation of the National Defense Program,* pt. 11, exhibit 411, pp. 4693-4721.

14. Ibid., exhibit 440, pp. 4677-92.

15. Quoted in ibid., at 4307.

16. Ibid., at 4345.

17. Ibid., at 4346.

18. Ibid.

19. Ibid., at 4261-82.

20. Ibid., at 4308.

21. *Christian Science Monitor,* March 28, 1942.

22. *Time,* April 6, 1942, pp. 15-16; *Newsweek,* April 6, 1942, pp. 46-48; Michael Straight, "Standard Oil: Axis Ally," *New Republic* 106 (April 6, 1942): 450-51; *PM:* April 1942.

23. *Investigation of the National Defense Program,* pt. 11, pp. 4283-4306.

24. For example, see *Christian Science Monitor,* March 28, 1942, and David Lawrence's syndicated column "Today in Washington," March 28, 1942.

25. See United States Senate, Committee on Banking and Currency, *Hearings on an Investigation into the Cause of Price Increases in Petroleum Products* (Washington: Government Printing Office, 1949); Anthony Sampson, *The Seven Sisters* (New York: Viking Press, 1975), pp. 126-29; Henrietta Larson, Evelyn Knowlton, and Charles Popple, *New Horizons 1927-1950* (New York: Harper & Row, 1971), pp. 672-73, 687-89.

26. Federal Trade Commission, *International Petroleum Cartel* (Washington: Government Printing Office, 1952).

27. *Time,* September 1, 1952, p. 69; *Newsweek,* September 1, 1952, pp. 49-53; *Fortune* (October 1952): 113-14; *United States* v. *Standard Oil Company (New Jersey) et al.,* U.S. District Court for the Southern District of New York, Civil Action No. 78-154.

28. *Newsweek,* September 1, 1952.
29. *Time,* September 1, 1952, p. 69.
30. *Fortune,* October 1952, p. 113.
31. Sampson, *Seven Sisters,* pp. 135-40.
32. Ibid., pp. 140-45.
33. Dean Acheson, *Present at the Creation* (New York: Signet, 1970), pp. 865-70.
34. United States Senate, Subcommittee on Multinational Corporations, *Multinational Corporations and United States Foreign Policy, Hearings,* 93d Cong., 1st and 2d sess. (Washington: Government Printing Office, 1974), pt. 8, appendix I, pp. 3-10.
35. Ibid., pp. 10-13.
36. Ibid., pt. 9, pp. 45-46; *United States v. Standard Oil Company,* D.D.C. 1953, Civil No. 1779-53; *Business Week,* January 17, 1953, p. 29.
37. Carl Solberg, *Oil Power* (New York: Mentor, 1976), pp. 196-98.
38. Sampson, *Seven Sisters,* pp. 155-59; John Blair, *The Control of Oil* (New York: Vintage, 1978), pp. 44-47.
39. *Multinational Corporations and United States Foreign Policy,* pt. 9, p. 46; Blair, *Control of Oil,* pp. 71-76.
40. Sampson, *Seven Sisters,* pp. 248-58.
41. Ibid., pp. 258-61; *Multinational Corporations and United States Foreign Policy,* pt. 9, p. 47.
42. Ibid.
43. Ibid., pp. 47-49.
44. Sampson, *Seven Sisters,* chap. 12.
45. *Multinational Corporations and United States Foreign Policy,* pt. 9, pp. 49, 51.
46. United States Senate, Subcommittee on Investigations, *Current Energy Shortage Oversight Series, Hearings,* 93d Cong., 2d sess. (Washington: Government Printing Office, 1974); *Multinational Corporations and United States Foreign Policy,* 9 parts.
47. Ibid., pt. 9, p. 49.
48. Ibid., pt. 9, p. 72.

Bibliography

MANUSCRIPTS

Bureau of Corporations Files. Record Group 122, National Archives.
Justice Department Files. Record Group 60, National Archives.
Charles J. Bonaparte Papers. Manuscript Division, Library of Congress.
James R. Garfield Papers. Manuscript Division, Library of Congress.
Philander C. Knox Papers. Manuscript Division, Library of Congress,
William H. Moody Papers. Manuscript Division, Library of Congress.
Richard Olney Papers. Manuscript Division, Library of Congress.
Theodore Roosevelt Papers. Manuscript Division, Library of Congress.
William H. Taft Papers. Manuscript Division, Library of Congress.
Bonaparte Family Papers. Maryland Historical Society, Baltimore.

GOVERNMENT DOCUMENTS

Arkansas. Attorney General's Office. *Biennial Report, 1911-1912.* Little
Rock: State Printing Office, 1912.
Iowa. Attorney General's Office. *Biennial Report, 1909-1910.* Des Moines:
Emory H. English, 1911.
Kansas. Attorney General's Office. *Biennial Report of the Attorney
General of Kansas.* 1905-1912. Topeka: State Printing Office,
1906-1913.
————.Legislature. *House Journal,* 14th biennial sess., 1905. Topeka:
State Printing Office, 1905.
————.*Senate Journal,* 14th biennial sess., 1905. Topeka: State Printing
Office, 1905.
Minnesota. Attorney General's Office. *Biennial Report, 1912-1913.*
Minneapolis: Syndicate Printing Company, 1913.
Mississippi. Attorney General's Office. *Biennial Report, 1907-1909.*
Nashville. Braden Printing Company, 1909.
Missouri. Attorney General's Office. *Missouri ex inf. Herbert S. Hadley v.
Standard Oil Company (of Indiana), Waters-Pierce Oil Company*

and Republic Oil Company. History of the Case, Issues Made by the Pleadings, Statement of Facts, Brief and Argument. 4 vols. Jefferson City, Mo.: H. Stephens Printing Company, 1907.

Nebraska. Attorney General's Office. *Report of the Attorney General, 1901–1902.* Lincoln: State Journal Company, 1902.

New York. Senate. *Report of the Committee on General Laws on the Investigation Relative to Trusts, 1888.* Troy, N.Y.: Troy Press Company, 1888.

——.State Assembly. *Proceedings of the Special Committee Appointed to Investigate Abuses in the Management of Railroads.* 8 vols. New York: Evening Post Steam Presses, 1879-1880.

Ohio. *Standard Oil Trust Cases, Record.* Columbus, 1899.

——.Attorney General's Office. *Annual Report.* 1899-1912. Columbus, 1899-1912.

——.Constitutional Convention. *Proceedings and Debates, 1912.* Columbus: F.J. Heer Publishing Company, 1922.

——.Senate. *Trust Investigation.* Columbus, 1899.

Texas. Attorney General's Office. *Report and Opinions of the Attorney General.* 1894-1909. Austin: Von Boeckman-Jones Company, 1894-1909.

——.Legislature. *House Journal.* 1889, 1901, 1907. Austin: Von Boeckman-Jones Company, 1889-1907.

——.Legislature. *Senate Journal.* 1889, 1901, 1907. Austin: Von Boeckman-Jones Company, 1889-1907.

——.Legislature. *Proceedings and Reports of the Bailey Investigation Committee.* Austin: Von Boeckman-Jones Company, 1907.

United States. Attorney General. *Annual Report.* 1891-1913, 1939-1941. Washington: Government Printing Office, 1891-1913, 1939-1941.

——.Bureau of Commerce. *Historical Statistics of the United States.* Washington: Government Printing Office, 1960.

——.Bureau of Corporations. *The International Harvester Company.* Washington: Government Printing Office, 1913.

——.Bureau of Corporations. *Report of the Commissioner of Corporations on the Transportation of Petroleum.* Washington: Government Printing Office, 1906.

——.Bureau of Corporations. *Report on the Petroleum Industry.* 2 vols. Washington: Government Printing Office, 1907.

——.Congress. *Congressional Record,* vols. 19-52, 50th Cong.-63d Cong. Washington: Government Printing Office, 1890-1914.

——.Congress. House of Representatives. Committee on Judiciary. *Hearings on House Bill 19745,* 60th Cong., 1st sess. Washington: Government Printing Office, 1908.

————.Congress. House of Representatives. Committee on Manufactures. *Investigation of Certain Trusts.* Washington: Government Printing Office, 1889.

————.Congress. Senate. *Executive Documents,* no. 604. 62d Cong., 2d sess. Washington: Government Printing Office, 1913.

————.Congress. Senate. Committee on Interstate Commerce. *Report . . . pursuant to Senate Resolution 98,* no. 1326. 62d Cong., 3d sess. Washington: Government Printing Office, 1913.

————.Congress. Senate. Committee on Investigating the National Defense Program. *Investigation of the National Defense Program.* 31 parts. 77thCong., 1st sess. Washington: Government Printing Office, 1941-1946.

————.Congress. Senate. Committee on Manufactures. *High Cost of Gasoline and Other Petroleum Products.* no. 1269. 67th Cong., 4th sess. Washington: Government Printing Office, 1923.

————.Congress. Senate. Committee on Patents. *Hearings before the Committee on Patents.* 77th Cong., 2d sess. Washington: Government Printing Office, 1942.

————.Congress. Senate. Committee on Privileges and Elections. *Campaign Contributions Testimony.* 2 vols. 62d Cong., 3d sess. Washington: Government Printing Office, 1913.

————.Congress. Senate. Subcommittee on Multinational Corporations, Committee on Foreign Relations. *Multinational Corporations and United States Foreign Policy. Hearings.* 93d Cong., 1st and 2d sess. Washington: Government Printing Office, 1974.

————.Congress. Senate. Subcommittee on Multinational Corporations, Committee on Foreign Relations. *The International Petroleum Cartel, the Iranian Consortium and U.S. National Security.* Washington: Government Printing Office, 1974.

————.Federal Trade Commission. *The Advance in the Price of Petroleum Products.* Washington: Government Printing Office, 1920.

————.Federal Trade Commission. *International Petroleum Cartel.* A Report to the Subcommittee on Monopoly, Select Committee on Small Business. 82d Cong., 2d sess. Washington: Government Printing Office, 1952.

————.Federal Trade Commission. *Petroleum Industry: Prices, Profits and Competition.* Senate Doc. 61. 70th Cong., 1st sess. Washington: Government Printing Office, 1928.

————.Federal Trade Commission. *Report on the Pacific Coast Petroleum*

Industry. Washington: Government Printing Office, 1921.

———.Federal Trade Commission. *Report on the Pipe Line Transportation of Petroleum.* Washington: Government Printing Office, 1916.

———.Federal Trade Commission. *Report on the Price of Gasoline in 1915.* Washington: Government Printing Office, 1917.

———.Federal Trade Commission. *Trust Laws and Unfair Competition.* Washington: Government Printing Office, 1916.

———.Fuel Administration. *Prices and Marketing Practices Covering the Distribution of Gasoline and Kerosene Throughout the United States.* By A.G. Maguire. Washington: Government Printing Office, 1919.

———.Industrial Commission. *Report of the Industrial Commission on Trusts and Industrial Combinations.* 13 vols. Washington: Government Printing Office, 1900.

———.Interstate Commerce Commission. *Report of the Investigation of Railroad Discriminations and Monopolies of Oil.* House Doc. 606. 59th Cong., 2d sess. Washington: Government Printing Office, 1907.

———.Laws, Statutes, etc. *Antitrust Laws with Amendments, 1890-1962.* Compiled by Gilman G. Udell. Washington: Government Printing Office, 1962.

———.Library of Congress. Legislative Reference Service. *Congress and the Monopoly Problem.* Washington: Government Printing Office, 1956.

United States v. *Standard Oil Company of New Jersey,* 173 Fed. 177 (1909); 221 U.S. 1 (1911). *Record.* 23 vols. Washington: Government Printing Office, 1909.

———.*Brief for Petitioner.* 3 vols. In the Circuit Court of the United States for the Eastern Division of the Eastern Judicial District of Missouri. Washington: Government Printing Office, 1909.

———.*Defendant's Brief.* 6 vols. In the Circuit Court of the United States for the Eastern Division of the Eastern Judicial District of Missouri. Washington: Government Printing Office, 1909.

———.*Brief for the United States.* No. 725. 2 vols. In the Supreme Court of the United States. Washington: Government Printing Office, 1910.

———.*Brief for the Standard Oil Company.* No. 725. 2 vols. In the Supreme Court of the United States. Washington: Government Printing Office, 1910.

NEWSPAPERS AND PERIODICALS

Business Week. 1953.
Chicago *Examiner.* 1914.
Christian Science Monitor. 1942.
Dallas Morning News. 1899, 1906, 1909.
Emporia (Kansas) *Gazette.* 1906.
Fortune. 1952, 1977.
The Independent. 1905-1911.
Literary Digest. 1892-1912.
Minneapolis Journal. 1910.
Minneapolis Morning Tribune. 1910.
National Petroleum News. 1910-1913.
New York American. 1903-1906.
New York Herald. 1897.
New York Press. 1905-1911.
New York *Sun.* 1905-1911.
New York Times. 1892-1913.
New York Tribune. 1898-1911.
New York *World.* 1892-1913.
Newsweek. 1942, 1952.
The Oil and Gas Journal. 1910-1912.
The Outlook. 1905-1911.
Philadelphia Press. 1907.
(New York) *PM.* 1942.
Raleigh *State Chronicle.* 1890.
Time. 1942, 1952.
Topeka State Journal. 1906.
Wall Street Journal. 1900-1911.
Washington *Times.* 1912.
Weekly Law Bulletin. 1899-1900.

CONTEMPORARY ARTICLES

Archbold, John D. "To the Press and Public." *New York Times,*
    September 26, 27, 1907.
———."The Standard Oil Company." *Saturday Evening Post* 180
    (December 7, 1907): 3-5.
Atwood, Albert W. "The Greatest Killing on Wall Street." *McClure's* 39
    (August 1912): 409-419.
Barde, F.S. "The Oil Fields and the Pipe Lines of Kansas." *The Outlook*
    80 (May 6, 1905): 19-32, 427-431.

Clark, Robert. "Breaking up a State Machine." *Cosmopolitan* 37 (October 1904): 665-670.

Dana, William F. "Monopoly under the National Anti-Trust Act." *Harvard Law Review* 7 (January 1894): 338-355.

Dodd, S.C.T. "The Present Legal Status of Trusts." *Harvard Law Review* 7 (October 1893): 157-169.

Ellis, Wade H. "The History of the Standard Oil Company in Ohio." *Ohio Magazine* 4 (January 1908): 1-10.

"Fifty Years of the Standard Oil Company." *World's Work* 39 (March 20, 1920): 431-432.

"The Foremost Democrat in Washington." *Current Literature* 40 (May 1906): 487-488.

Forrest, J. D. "Anti-Monopoly Legislation in the United States." *American Journal of Sociology* 1 (January 1896): 411-425.

Gatlin, Dana. "'What I Am Trying To Do': An Interview with Hon. W. R. Stubbs." *World's Work* 24 (May 1912): 59-67.

"Good News for Oil and Steel Workers." *The Outlook* 118 (April 17, 1918): 612.

Goodnow, Frank J. "Trade Combinations at Common Law." *Political Science Quarterly* 12 (June 1897): 212-245.

Harger, Charles M. "Kansas' Battle for Its Oil Interests." *American Monthly Review of Reviews* 31 (April 1905): 471-474.

Haskell, H. J. "The People His Clients." *The Outlook* 88 (March 28, 1908): 417-419.

Hoch, Edward W. "Kansas and the Standard Oil Company." *The Independent* 58 (March 2, 1905): 461-463.

J. E. ———. "The History of the Standard Oil Letters." *Hearst's Magazine* 21 (May-June 1912): 2204-2216, 2362-2376.

Janeway, Eliot. "Economic Warfare." *Fortune* 24 (August 1942): 101, 118-122.

Johnson, Charles. "The Attorney-General and the Trusts." *Harper's Weekly* 55 (May 22, 1911): 8.

Kellogg, Frank B. "Results of the Standard Oil Decision." *American Review of Reviews* 45 (June 1912): 728-730.

Kinkead, Edgar B. "A Sketch of the Supreme Court of Ohio." *The Green Bag* 7 (March 1895): 273-294.

Lloyd, Henry D. "The Story of a Great Monopoly." *Atlantic Monthly* 47 (March 1881): 317-334.

Lockwood, Frank C. "Governor Hadley of Missouri." *The Independent* 46 (April 8, 1909): 742-746.

Lowry, Edward G. "The Supreme Court Speaks." *Harper's Weekly* 55 (June 3, 1911): 8.

McLaurin, John J. "The Oil Situation in Kansas." *The Outlook* 80 (June 17, 1905): 427-431.

Marcosson, Isaac F. "The Kansas Oil Fight." *World's Work* 10 (May 1905): 6155-6166.

Monnett, Frank S. "Bryan and the Trusts: An Anti-Trust View." *American Review of Reviews* 22 (1900): 439-443.

———. "Transportation Franchises Always the Property of Sovereignty." *Arena* 26 (August 1901): 113-127.

Morse, Sherman. "The Taming of Rogers." *American Magazine* 62 (June 1906): 93-96.

"New Lights of Asia a General Convenience." *Literary Digest* 55 (November 17, 1917): 79.

Phillips, David Graham. "The Treason of the Senate." *Cosmopolitan Magazine* 41 (July 1906): 267-276.

"Pioneers of Standard Oil." *Nation* 103 (December 14, 1916): 556.

"President Taft on the Trusts." *The Outlook* 99 (September 30, 1911): 249-250.

Reiman, Guenter. "Standard Oil and I. G. Farben." *New Republic* 105 (August 4, 1941): 147-149.

Rogers, Henry H. "Interview." *New York Times,* January 7, 9, 10, 11, 13, 14, 1906.

Sayers, Joseph D. "Anti-Trust Legislation." *North American Review* 169 (August 1899): 210-217.

"Standard Oil and China, Partners." *Literary Digest* 48 (February 28, 1914): 421-422.

"Standard Oil and the Eight Hour Day." *The Outlook* 111 (September 15, 1915): 109-110.

"The Standard Oil Company at the Bar." *Arena* 35 (March 1906): 307-310.

Straight, Michael. "Standard Oil: Axis Ally." *New Republic* 106 (April 6, 1942): 450-451.

Tarbell, Ida M. "Kansas and the Standard Oil Company." *McClure's* 25 (September-October 1905): 469-481, 608-622.

"At Thirty-Nine He Became a Great Captain of Industry." *Current Opinion* 64 (January 1918): 21-22.

"Those Who Can Afford Welfare Work." *World's Work* 36 (June 1918): 131-132.

"Trust Busting vs. Regulation." *The Outlook* 104 (August 2, 1913): 731-732.

"What's the Explanation?" *Everybody's Magazine* 26 (June 1912): 817-818.

White, William Allen. "Political Signs of Promise." *The Outlook* 80 (July 15, 1905): 667-670.

Wickersham, George. "Recent Interpretation of the Sherman Act." *Michigan Law Review* 10 (November 1911): 1-25.

Wormser, I. Maurice. "Piercing the Veil of Corporate Entity." *Columbia Law Review* 12 (June 1912): 496-518.

Yale, E. S. "Carriers of Light and Laughter." *World Outlook* 3 (November 1917): 14-15.

OTHER PUBLISHED PRIMARY SOURCES

Adams, Frederick U. *The Waters Pierce Case in Texas, Compiled from the Series of Press Articles Entitled: Battling with a Great Corporation.* St. Louis: Skinner and Kennedy, 1908.

Baker, Ray Stannard, and Dodd, William E., eds. *The Public Papers of Woodrow Wilson.* 6 vols. New York: Harper and Brothers, 1925-1927.

Bryce, James. *The American Commonwealth.* 2 vols. New York: Macmillan Company, 1888.

Butt, Archibald W. *Taft and Roosevelt: The Intimate Letters of Archie Butt.* 2 vols. Garden City, N.Y.: Doubleday, Doran, 1930.

Civic Federation of Chicago. *Chicago Conference on Trusts.* Chicago: Civic Federation of Chicago, 1900.

Cocke, William A. *The Bailey Controversy in Texas.* 2 vols. San Antonio, Texas: Cocke Company, 1908.

Cook William W. *Trusts: The Recent Combinations in Trade.* New York: L. K. Strouse and Company, 1888.

Cotner, Robert C., ed. *Addresses and State Papers of James S. Hogg.* Austin: University of Texas Press, 1951.

Crawford, W. L. *Crawford on Baileyism.* Dallas: Eclectic News Bureau, 1907.

Danelski, David J., and Tulchin, Joseph S., eds. *The Autobiographical Notes of Charles Evans Hughes.* Cambridge: Harvard University Press, 1973.

Dodd, Samuel C. T. *Memoirs of S. C. T. Dodd.* New York: R. G. Cook, 1907.

Eliot, Charles William. *American Contributors to Civilization.* New York: The Century Company, 1898.

Finty, Jr., Tom. *Anti-Trust Legislation in Texas.* Galveston: A. H. Belo & Company, 1916.

Foraker, Joseph B. *Notes of a Busy Life.* 2 vols. Cincinnati: Stewart and Kidd Company, 1916.

Gray, W. H. *The Rule of Reason in Texas.* Houston: W. H. Gray, 1912.

Howe, Mark DeWolfe, ed. *Holmes-Pollock Letters.* 2 vols. Cambridge: Harvard University Press, 1941.

Hughes, Charles Evans. *The Supreme Court of the United States.* New York: Garden City Publishing Company, 1928.

Kellogg, Frank B. *Standard Oil Company of New Jersey v. United States . . . Oral Argument of Frank B. Kellogg on behalf of the United States.* Washington: Government Printing Office, 1911.

Lloyd, Henry D. *Wealth Against Commonwealth.* Washington: National Home Library Association, 1936.

McKee, Thomas H. *The National Conventions and Platforms of All Political Parties.* Baltimore: Friedwald Company, 1906.

Milburn, John G. *Standard Oil Company of New Jersey v. United States . . . Oral Arguments on Behalf of Appellants.* New York: C. G. Burgoyne, 1911.

Montague, Gilbert H. *The Rise and Progress of Standard Oil.* New York: Harper and Brothers, 1904.

Morison, Elting E., ed. *The Letters of Theodore Roosevelt.* 8 vols. Cambridge: Harvard University Press, 1951-1954.

National Civic Federation. *Proceedings of the National Conference on Trusts and Combinations.* New York: National Civic Federation, 1908.

Oil City Derrick. *Pure Oil Trust v. Standard Oil Company.* Oil City, Pa.: Derrick Publishing Company, 1901.

Richardson, James D., comp. *The Messages and Papers of Presidents.* 20 vols. New York: Bureau of National Literature, 1903-1912.

Rockefeller, John D. *Random Reminiscences of Men and Events.* New York: Doubleday, Page & Company, 1909.

Roosevelt, Theodore. *An Autobiography.* New York: Charles Scribner's Sons, 1920.

———. *The Works of Theodore Roosevelt.* 20 vols. New York: Charles Scribner's Sons, 1926.

Senter, E. G. *The Bailey Case Boiled Down.* Dallas: Flag Publishing Company, 1908.

Shoemaker, Floyd C., and Guiter, Sarah, eds. *Messages and Proclamations of the Governors of the State of Missouri.* 16 vols. Columbia, Mo.: State Historical Society, 1928.

Taft, William H. *The Anti Trust Act and the Supreme Court.* New York: Harper and Brothers, 1914.

Tarbell, Ida M. *All in the Day's Work.* New York: Macmillan Company, 1939.

————. *The History of the Standard Oil Company.* 2 vols. New York: McClure, Phillips & Company, 1904.

Watson, David K. *The Constitution of the United States.* 2 vols. Chicago: Callaghan and Company, 1910.

————. *History of American Coinage.* New York: G. P. Putnam's Sons, 1899.

Wells, David Ames. *Recent Economic Changes in the United States.* New York: D. Appleton, 1889.

Wickersham, George. *The Changing Order.* New York: G. P. Putnam's Sons, 1914.

Wilson, Woodrow. *The New Freedom.* New York: Doubleday, 1913.

Winkler, Ernest R. *Platforms of Political Parties in Texas.* Austin: University of Texas Press, 1916.

DISSERTATIONS AND THESES

Andreano, Ralph. "The Emergence of New Competition in the American Petroleum Industry before 1911." Ph.D. dissertation, Northwestern University, 1960.

Dudden, Arthur P. "Antimonopolism, 1865-1890." Ph.D. dissertation, University of Michigan, 1950.

Feller, John Q. "Theodore Roosevelt, the Department of Justice, and the Trust Problem." Ph.D. dissertation, Catholic University, 1968.

German, James C. "Taft's Attorney General: George W. Wickersham." Ph.D. dissertation, New York University, 1969.

Holcomb, Bob Charles. "Senator Joe Bailey, Two Decades of Controversy." Ph.D. dissertation, Texas Technological College, 1968.

Leinwand, Gerald. "A History of the United States Bureau of Corporations" Ph.D. dissertation, New York University, 1962.

Rumble, Allen W. "Rectitude and Reform: Charles J. Bonaparte and the Politics of Gentility, 1851-1921." Ph.D. dissertation, University of Maryland, 1971.

Sheifer, Isobel C. "Ida M. Tarbell and Morality in Big Business." Ph.D. dissertation, New York University, 1967.

Wagner, Robert L. "The Gubernatorial Career of Charles Allen Culberson." Master's thesis, University of Texas, 1954.

Worner, Lloyd E. "The Public Career of Herbert S. Hadley," Ph.D. dissertation, University of Missouri, 1946.

PUBLISHED SECONDARY SOURCES

Abels, Jules. *The Rockefeller Billions.* New York: Macmillan Company, 1965.

Acheson, Dean. *Present at the Creation*. New York: Signet, 1974.

Acheson, Sam H. *Joe Bailey, the Last Democrat*. New York: Macmillan Company, 1932.

———. *35,000 Days in Texas*. New York: Macmillan Company, 1938.

Adelman, M. A. "Effective Competition and the Antitrust Laws." *Harvard Law Review* 61 (September 1948): 1289-1350.

———. *The World Petroleum Market*. Baltimore: Johns Hopkins Press, 1972.

Allen, Ruth A. *The Great Southwest Strike*. Austin: University of Texas Press, 1942.

Anderson, Donald F. *William Howard Taft*. Ithaca: Cornell University Press, 1968.

Barr, Alwyn, *Reconstruction to Reform: Texas Politics, 1876-1906*. Austin: University of Texas Press, 1971.

Beard, Charles A., and Beard, Mary R. *The Rise of American Civilization*. 2 vols. New York: Macmillan Company, 1927.

Beth, Loren P. "Justice Harlan and the Uses of Dissent." *American Political Science Review* 49 (December 1955): 1080-1104.

Bishop, Joseph B. *Charles J. Bonaparte*. New York: Charles Scribner's Sons, 1922.

———. *Theodore Roosevelt and His Time*. 2 vols. New York: Charles Scribner's Sons, 1920.

Blair, John M. *The Control of Oil*. New York: Vintage Books, 1978.

Blum, John M. *The Republican Roosevelt*. Cambridge: Harvard University Press, 1954.

Bonbright, James C., and Means, Gardiner C. "Holding Companies." *Encyclopedia of the Social Sciences,* 9:403-404. New York: Macmillan Company, 1937.

Bright, John D., et al. *Kansas: The First Century*. 4 vols. New York: 1956.

Bryn-Jones, David. *Frank B. Kellogg*. New York: G. P. Putnam's Sons, 1937.

Burns, Arthur Robert. *The Decline of Competition*. New York: McGraw-Hill, 1936.

Carr, Albert H. *John D. Rockefeller's Secret Weapon*. New York: McGraw-Hill, 1962.

Clark, Floyd B. *The Constitutional Doctrines of Justice Harlan*. Baltimore: Johns Hopkins University Press, 1915.

Clark, John D. *The Federal Trust Policy*. Baltimore: Johns Hopkins University Press, 1931.

Coletta, Paola E. *The Presidency of William Howard Taft*. Lawrence: University of Kansas Press, 1973.

Collier, Peter, and Horowitz, David. *The Rockefellers.* New York: Signet, 1977.

Connelley, William E. *A Standard History of Kansas and Kansans.* 5 vols. Chicago, 1918.

Cook, Roy C. *Control of the Petroleum Industry By Major Oil Companies,* Temporary National Economic Committee monograph 39. Washington: Government Printing Office, 1941.

Cotner, Robert C. *James Stephen Hogg.* Austin: University of Texas Press, 1959.

Croly, Herbert. *Life of Mark Hanna.* New York: Macmillan Company, 1923.

Cummings, Homer, and McFarland, Carl. *Federal Justice.* New York: Macmillan Company, 1937.

Davis, Joseph S. *Essays in the Earlier History of American Corporations.* Cambridge: Harvard University Press, 1917.

De Shields, James T. *They Sat in High Places.* San Antonio: Naylor Company, 1940.

Destler, Chester M. *Henry Demarest Lloyd and the Empire of Reform.* Philadelphia: University of Philadelphia Press, 1963.

Dewey, Donald. *Monopoly in Economics and Law.* Chicago: Rand McNally & Company, 1959.

Dictionary of American Biography. 22 vols. New York: Charles Scribner's Sons, 1928-1936.

Dill, James B. *The Statute and Case Law of New Jersey Relating to Business Companies.* Camden, N.J.: Press of S. Chew and Sons Company, 1910.

Dishman, Robert B. "Mr. Justice White and the Rule of Reason." *Review of Politics* 12 (1951): 229-232.

Dorfman, Joseph. *The Economic Mind in American Civilization.* New York: Viking Press, 1949.

Douglas, Richard L. "A History of Manufactures in the Kansas District." *Collections of the Kansas State Historical Society* 11. Topeka: State Printing Office, 1910, pp. 193-203.

Dunham, Allison, and Kurland, Philip B., eds. *Mr. Justice.* Chicago: University of Chicago Press, 1962.

Engler, Robert. *The Politics of Oil.* Chicago: University of Chicago Press, 1961.

Flynn, John T. *God's Gold.* New York: Harcourt, Brace and Company, 1932.

Fosdick, Raymond B. *John D. Rockefeller, Jr.: A Portrait.* New York: Harper and Brothers, 1956.

Friedman, Leon, and Israel, Fred L., eds. *Justices of the United States Supreme Court.* 4 vols. New York: R. R. Bowker Company, 1969.

Galambos, Louis. *The Public Image of Big Business in America, 1880-1940.* Baltimore: Johns Hopkins University Press, 1975.

Geiger, Louis G. *Joseph W. Folk.* Columbia, Mo.: University of Missouri Press, 1953.

Gibb, George S., and Knowlton, Evelyn H. *The Resurgent Years: 1911-1927.* New York: Harper and Brothers, 1956.

Giddens, Paul H. *Standard Oil Company (Indiana).* New York: Appleton-Century-Crofts, 1955.

Goldman, Eric F. *Charles J. Bonaparte, Patrician Reformer: His Earlier Career.* Baltimore: Johns Hopkins University Press, 1943.

Hamilton, Walton. *The Pattern of Competition.* New York: Columbia University Press, 1940.

Hamilton, Walton, and Till, Irene. *Anti Trust in Action.* Temporary National Economic Committee monograph 16. Washington: Government Printing Office, 1940.

Handler, Milton. *A Study of the Construction and Enforcement of Federal Antitrust Law.* Temporary National Economic Committee monograph 38. Washington: Government Printing Office, 1941.

Harbaugh, William H. *The Life and Times of Theodore Roosevelt.* New York: Collier Books, 1963.

Harvard Graduate School of Business Administration. *Oil's First Century.* Cambridge: Harvard Graduate School of Business Administration, 1960.

Hawley, Ellis. *The New Deal and the Problem of Monopoly.* Princeton: Princeton University Press, 1964.

Hidy, Ralph, and Hidy, Muriel. *Pioneering in Big Business.* New York: Harper and Brothers, 1955.

Hofstader, Richard. *The Paranoid Style in American Politics.* New York: Random House, 1964.

Howe, Mark DeWolfe. *Justice Oliver Wendell Holmes.* 2 vols. Cambridge: Harvard University Press, 1957-1962.

Ise, John. *United States Oil Policy.* New Haven: Yale University Press, 1926.

Jacobson, Charles. *The Life of Jeff Davis.* Little Rock: Parke-Harper Publishing Company, 1925.

Jencks, Jeremiah, W., and Clark, W. E. *The Trust Problem.* 5th ed. Garden City, N.Y.: Doubleday, Doran and Company, 1929.

Johnson, Arthur M. "Antitrust Policy in Transition, 1908." *Mississippi Valley Historical Review* 48 (1961): 415-434.

———. *The Development of American Petroleum Pipelines.* Ithaca: Cornell University Press, 1956.

———. *Petroleum Pipelines and Public Policy, 1906-1959.* Cambridge: Harvard University Press, 1967.

———. "Theodore Roosevelt and the Bureau of Corporations." *Mississippi Valley Historical Review* 45 (March 1959): 571-590.

Jones, Eliot. *The Trust Problem in the United States.* New York: Macmillan Company, 1926.

Jordan, Philip D. *Ohio Comes of Age.* Columbus: Ohio State Archaeological and Historical Society, 1943.

Josephson, Matthew. *The Polticos.* New York: Harcourt, Brace, 1938.

———. *The Robber Barons.* New York: Harcourt, Brace, 1934.

Kemnitzer, William J. *The Rebirth of Monopoly: A Critical Analysis of Economic Conduct in the Petroleum Business.* New York: Harper and Brothers, 1938.

Klinkhamer, Marie C. *Edward Douglas White.* Washington, 1943.

Kolko, Gabriel. "American Business and Germany, 1930-1941." *Western Political Quarterly* 15 (December 1962): 713-728.

———. *Triumph of Conservatism.* Chicago: Quadrangle Books, 1967.

Konefsky, Samuel J. *The Legacy of Holmes and Brandeis.* New York: Macmillan Company, 1956.

La Follette, Belle C., and La Follette, Fola. *Robert M. LaFollette.* 2 vols. New York: Macmillan Company, 1953.

Lamar, Clarinda P. *The Life of Joseph Rucker Lamar.* New York: G. P. Putnam's Sons, 1926.

Larson, Henrietta; Knowlton, Evelyn; and Popple, Charles. *New Horizons, 1927-1950.* New York: Harper & Row, 1971.

Ledbetter, Cal. "Jeff Davis and the Politics of Combat." *Arkansas Historical Quarterly* 33 (Spring 1974): 15-37.

Letwin, William. *Law and Economic Policy in America.* New York: Random House, 1965.

Link, Arthur S. *Wilson: The New Freedom.* Princeton: Princeton University Press, 1956.

———. *Wilson: The Road to the White House.* Princeton: Princeton University Press, 1947.

———. *Woodrow Wilson and the Progressive Era.* New York: Harper Torchbook ed., 1963.

Lloyd, Caroline A. *Henry D. Lloyd.* New York: G. P. Putnam's Sons, 1912.

Long, Hazel T. "Attorney General Herbert S. Hadley versus the Standard Oil Trust." *Missouri Historical Review* 35 (1940): 171-187.

McDevitt, Matthew. *Joseph McKenna.* Washington, 1946.

McGee, John S. "Predatory Price Cutting: The Standard Oil (N.J.) Case." *Journal of Law and Economics* 1 (1958): 137-169.

McLean, Joseph E. *William Rufus Day.* Baltimore: Johns Hopkins University Press, 1946.

Marcosson, Isaac F. *Black Golconda.* New York: Harper and Brothers, 1924.

Marshall, Carrington T. *A History of Courts and Lawyers of Ohio.* New York: American Historical Society, 1934.

Martin, Edward S. *Life of Joseph Hodges Choate.* 2 vols. New York: Charles Scribner's Sons, 1920.

Mason, Alpheus T. *Louis D. Brandeis: A Free Man's Life.* New York: Viking Press, 1946.

———. *William Howard Taft: Chief Justice.* London: Oldbourne Book Company, 1964.

Meyer, Balthasar H. *History of the Northern Securities Case.* 1906; Reprint, New York: Da Capo Press, 1972.

Meyers, Allen O. *Bosses and Boodle in Ohio Politics.* Cincinnati: Lyceum Publishing Company, 1895.

Miller, Arthur S. *The Supreme Court and American Capitalism.* New York: Free Press, 1968.

Moore, Austin L. *John D. Archbold and the Early Development of Standard Oil.* New York: Macmillan Company, 1930.

Morison, Samuel E., and Commager, Henry Steele. *The Growth of the American Republic.* 4th ed. 2 vols. New York, 1951.

Mowry, George E. *The Era of Theodore Roosevelt.* New York: Harper and Brothers, 1958.

Nash, Gerald D. *United States Oil Policy, 1890-1964.* Pittsburgh: University of Pittsburgh Press, 1968.

National Cyclopaedia of American Biography. 37 vols. New York: James T. White and Company, 1898-1951.

Nevins, Allen. *John D. Rockefeller: The Heroic Age of American Enterprise.* 2 vols. New York: Charles Scribner's Sons, 1940.

———. *Study in Power: John D. Rockefeller.* 2 vols. New York: Charles Scribner's Sons, 1953.

Oliphant, Herman. *Cases on Trade Regulation Selected from Decisions of English and American Courts.* St. Paul: West Publishing Company, 1923.

Paul, Arnold. *Conservative Crisis and the Rule of Law.* Rev. ed. New York: Harper and Row, 1969.

Penrose, Edith. *The International Petroleum Industry.* London: Allen and Unwin, 1968.

Pringle, Henry F. *The Life and Times of William Howard Taft.* 2 vols. New York: Toronto, Farrer & Rinehart, 1939.

———. *Theodore Roosevelt.* 2 vols. New York: Harcourt, Brace, 1931.

Purdy, Harry L.; Lindahl, Martin; and Carter, William. *Corporate Concentration and Public Policy.* 2d ed. New York: Prentice-Hall, 1950.

Pusey, Merlo. *Charles Evans Hughes.* 2 vols. New York: Macmillan Company, 1951.

Ralston, Jackson H. *Study and Report for the American Federation of Labor upon Judicial Control over Legislatures as to Constitutional Questions.* 2d ed. Washington: Law Reporter Printing Company, 1923.

Reed, George I., ed. *Bench and Bar of Ohio.* 4 vols. Chicago: Century Publishing Company, 1897.

Richardson, Rupert N. *Colonel Edward M. House: The Texas Years, 1858-1912.* Abilene: Harden-Simmons University Publications in History, 1964.

———. *Texas: The Lone Star State.* New York: Prentice-Hall, 1943.

Rister, Carl Coke. *Oil! Titan of the Southwest.* Norman: University of Oklahoma Press, 1949.

"The Rule of Reason in Loose-Knit Combinations." *Columbia Law Review* 22 (February 1932): 291-324.

Sait, Edward M. *American Parties and Elections,* New York: The Century Company, 1927.

Sampson, Anthony. *The Seven Sisters.* New York: Viking Press, 1975.

Seager, Henry R., and Gulick, Charles A. *Trust and Corporation Problems.* New York: Harper and Brothers, 1929.

Smith, Ralph A. "The Farmers' Alliance in Texas, 1875-1900." *Southwestern Historical Quarterly* 48 (January 1945): 346-369.

Solberg, Carl. *Oil Power.* New York: Mentor, 1976.

Stewart, Frank M., and Clark, Joseph L. *The Constitution and Government of Texas.* Rev. ed. Boston: D. C. Heath and Company, 1933.

Stocking, George W. *The Oil Industry and the Competitive System.* Boston: Houghton Mifflin Company, 1925.

Stocking, George W. et al. *Cartels in Action: Case Studies in International Business Diplomacy.* New York: Twentieth Century Fund, 1947.

Stocking, George W., and Watkins, Myron W. *Cartels or Competition?* New York: Twentieth Century Fund, 1948.

Stromberg, Roland. "American Business and the Approach of War, 1935-1941." *Journal of Economic History* 13 (Winter 1953): 58-78.

Strong, Theron G. *Joseph H. Choate.* New York: Dodd, Mead and Company, 1917.

Tarbell, Ida M. *The Nationalizing of Business*. New York: Macmillan Company, 1936.

Tennant, Richard B. *The American Cigarette Industry*. New Haven: Yale University Press, 1950.

Thorelli, Hans B. *Federal Antitrust Policy*. Baltimore: Johns Hopkins University Press, 1955.

Twiss, Benjamin. *Lawyers and the Constitution*. Princeton: Princeton University Press, 1942.

Walker, Albert H. *A History of the Sherman Act*. New York: Equity Press, 1910.

Walters, Everett. *Joseph Benson Foraker*. Columbus: Ohio History Press, 1948.

Warner, Hoyt Landon. *Progressivism in Ohio*. Columbus: Ohio State University Press, 1964.

Watkins, Myron W. *Industrial Combinations and Public Policy*. Boston: Houghton Mifflin Company, 1927.

Webb, Walter Prescott, ed. *The Handbook of Texas*. 2 vols. Austin: Texas State Historical Association, 1952.

Weisenberger, E. P., and Roseboom, E. H. *A History of Ohio*. New York: Prentice-Hall, 1934.

Wesser, Robert F. *Charles Evans Hughes: Politics and Reform in New York, 1905-1910*. Ithaca: Cornell University Press, 1967.

White, Gerald T. *Formative Years in the Far West: Standard Oil of California*. New York: Appleton-Century-Crofts, 1962.

Whitney, Simon N. *Antitrust Politics*. 2 vols. New York: Twentieth Century Fund, 1958.

Who's Who in America, 1908-1909. Edited by Albert N. Marquis. Chicago: A. N. Marquis and Company, 1909.

Who's Who in Jurisprudence. Edited by John William Leonard. Brooklyn: A. W. Steven Printing Company, 1925.

Wiebe, Robert H. *Businessmen and Reform*. Chicago: Quadrangle Press, 1962.

———. "The House of Morgan and the Executive." *American Historical Review* 65 (October 1959): 49-60.

———. *The Search for Order, 1877-1920*. New York: Hill and Wang, 1967.

Williamson, Harold F. et al. *The American Petroleum Industry: The Age of Energy*. Evanston: Northwestern University Press, 1963.

Woodward, C. Vann. *Origins of the New South*. Baton Rouge: Louisiana State University Press, 1951.

Zornow, William F. *Kansas: A History of the Jayhawk State*. Norman: University of Oklahoma Press, 1957.

TABLE OF CASES

Addystone Pipe & Steel Company v. *United States,* 175 U.S. 211 (1899).

Anderson et al. v. *United States,* 171 U.S. 604 (1899).

Baker v. *Grice,* 169 U.S. 284 (1898).

Central Ohio Salt Company v. *Guthrie,* 35 Ohio 666 (1880).

Collins v. *Locke,* 4 L.R.A.C. 674 (1879).

John Dyer's Case, Y.B. 2 Hen. V, f. 5, pl. 26 (1415).

Hathaway v. *State,* 36 Tex. Crim. Rep. 261 (1896).

Hopkins et al. v. *United States,* 171 U.S. 578 (1898).

In re Grice, 79 Fed. 627 (1897).

Mallory v. *Hanaur Oil-Works,* 88 Tenn. 598 (1888).

Missouri ex inf. Hadley v. *Standard Oil Company of Indiana, et al.,* 194 Mo. 124 (1906).

Missouri ex inf. Hadley v. *Standard Oil Company, et al.,* 218 Mo. 1 (1909).

Missouri ex inf. Hadley v. *Standard Oil Company, et al.,* 251 Mo. 271 (1913).

Mitchell v. *Reynolds,* 1 P. Wms. 181, 24 Eng. Rep. 347 (1711).

Mogul Steamship Company v. *McGregor,* 21 Q.B. Div. 554 (1888).

Nordenfelt v. *Maxim Nordenfelt Gun Company,* App. Cas. 535 (1834).

Northern Securities Company v. *United States,* 193 U.S. 197 (1904).

Ohio v. *Standard Oil Company of Ohio,* 49 Ohio 137 (1892).

Ohio ex rel. Monnett v. *Buckeye Pipe Line Company, etc.* 61 Ohio 520 (1900).

Ohio ex rel. Wachenheimer v. *Standard Oil Company of Ohio, et al.,* 15 Ohio Cir. Ct. Rep. (New Series) 212 (1907).

People v. *North Sugar Refining Company,* 121 N.Y. 582 (1890).

Rice v. *Standard Oil Co.,* 134 Fed. 464 (1905).

Richardson v. *Buhl,* 77 Mich. 632 (1889).

Standard Oil Company, et al. v. *State,* 117 Tenn. 618 (1907).

Standard Oil Company of Indiana v. *Missouri ex inf. Hadley; Republic Oil Company* v. *Missouri ex inf. Hadley,* 224 U.S. 270 (1912).

Standard Oil Company of Kentucky v. *Tennessee,* 217 U.S. 413 (1910).

Standard Oil Company of New Jersey v. *United States,* 221 U.S. 1 (1911).

State v. *Kelly,* 71 Kan. 811 (1905).

State v. *Nebraska Distilling Company,* 29 Neb. 700 (1890).

State v. *Railway Company,* 40 Ohio 504 (1884).

State v. *Standard Oil Company,* 150 Iowa 46 (1911).

State ex rel. Edward T. Young v. *Standard Oil Company,* 111 Minn. 85 (1910).

Tennessee ex rel. Cates v. *Standard Oil Company of Kentucky,* 120 Tenn. 86 (1908).

Texas ex rel. Attorney General v. *Waters-Pierce Oil Company,* 67 S.W. 1057 (1902).

United States v. *Addystone Pipe & Steel Company,* 85 Fed. 271 (1898).

United States v. *American Tobacco,* 221 U.S. 106 (1911).

United States v. *E. C. Knight Company,* 156 U.S. 1 (1895).

United States v. *International Harvester Company,* 274 U.S. 693 (1927).

United States v. *Joint-Traffic Association,* 171 U.S. 505 (1898).

United States v. *Northern Securities Company,* 120 Fed. 721 (1903).

United States v. *Standard Oil Company of Indiana,* 155 Fed. 305 (1907).

United States v. *Standard Oil Company of New Jersey,* 152 Fed. 290 (1907).

United States v. *Standard Oil Company of New Jersey,* 173 Fed. 177 (1909).

United States v. *Trans-Missouri Freight Association,* 166 U.S. 290 (1897).

United States v. *United Shoe Machinery Company,* 247 U.S. 32 (1918).

United States v. *United States Steel Corporation,* 251 U.S. 417 (1920).

Waters-Pierce Oil Company v. *Texas,* 19 Tex. Civ. App. 1 (1898).

Waters-Pierce Oil Company v. *Texas,* 177 U.S. 28 (1900).

Waters-Pierce Oil Company v. *Texas,* 48 Tex. Civ. App. 162 (1907).

Waters-Pierce Oil Company v. *Texas,* 212 U.S. 86 (1909).

Index

About the Author

Bruce Bringhurst, a staff member of the Claremont Colleges, is a specialist in twentieth-century legal and constitutional history. He has published in the *Southern California Quarterly.*